NON-CIRCULATING MATERIAL

DISCARDED

Developments in Fluidization and Fluid Particle Systems

Elmer, L. Gaden, Jr., Series Editor

John C. Chen, Volume Editor

Hamid Arastoopour, L.S. Fan, Desmond King, Alan W. Weimer and Wen-Ching Yang, Volume Co-Editors

S. Altobelli	L.G. Gibilaro	H.Y. Pan
S.C. Arnold	D. Gidaspow	Anand Prakash
E. Aksoy	R. Gudhe	K.R. Rajagopal
H.E. Barner	Rich Harner	James Radar
G. Gregory Benge	Lu Huilin	S. Rapagna
Sidney Benson	Y. Itani	J.C. Schouten
J.X. Bouillard	George E. Klinzing	O.C. Snip
Ray Chrisman	H.O. Kono	Arthur M. Squires
John Cleveland	R. Korbee	J.J. Su
Ray Cocco	E. Koresawa	Steve Tallon
Clive E. Davis	Juan Llibre	T.T. Tsotsis
C. Ercan	M. Massoudi	C.M. van den Bleek
Liang-Shih Fan	R.G. Minet	H.-Y. Xie
P.U. Foscolo	M. Mortensen	Yan Xue
D. Geldart		R.C. Yalamanchili

AIChE Staff
Maura N. Mullen, Managing Editor; Julie A. McBride, Editorial Assistant
Cover Design: Joseph A. Roseti

Inquiries regarding the publication of Symposium Series Volumes should be directed to:
Mark Rosenzweig, Editor-in-Chief
American Institute of Chemical Engineers, 345 E 47 St., New York, N.Y. 10017
(212) 705-7576 • FAX: (212) 705-7812

AIChE Symposium Series

1995

American Institute of Chemical Engineers

© 1995
American Institute of Chemical Engineers (AIChE)
345 E 47 Street
New York. N.Y. 10017

AIChE shall not be responsible for statements or opinions advanced in their papers or printed in their publications.

ISBN 0-8169-0688-2

ISSN 0065-8812

All rights reserved whether the whole or part of the material is concerned, specifically those of translation, reprinting, re-use of illustrations, broadcasting, electronic networks, reproduction by photocopying machine or similar means, and storage of data in banks.

Authorization to photocopy items for internal use, or the internal or personal use of specific clients, is granted by AIChE for libraries and other users registered with the Copyright Clearance Center Inc., provided that the $3.50 per copy is paid directly to CCC, 222 Rosewood Drive, Danvers, MA 01923. This consent does not extend to copying for general distribution, for advertising, or promotional purposes, for inclusion in a publication, or for resale.

Articles published before 1978 are subject to the same copyright conditions and the fee is $3.50 for each article. AIChE Symposium Series fee code: 0065-8812/1995.

Foreword

This volume of the AIChE Symposium Series continues the tradition of an annual volume presenting recent developments in fluidization and fluid-particle systems. The papers are selected by peer reviews from those presented in ten sessions at the AIChE Annual Meeting in San Francisco, California, November 13, through 18, 1994.

This year's volume includes an invited plenary paper by Professor L.S. Fan entitled "Bubble Dynamics in Liquid-Solid Suspensions". This is a substantial review of state-of-science in this area of fluid-particle systems. In addition, fifteen other papers present recent developments concerning bubbling fluidized beds, circulating fluidized beds, vibrating beds, granular and suspension two-phase flows, characterization of powders, and novel experimental techniques. The developments are pertinent to such applications as particle processing, particle flows, fluidized reactors, and fluidized combustors.

I wish to acknowledge the fine work of the chair/co-chairs of the sessions for their collection and selection of papers, and the reviewers whose thoughtful comments provided guidance for the publication of this volume.

John C. Chen, *Volume Editor*
Carl R. Anderson Professor of
 Chemical Engineering
Lehigh University
Bethlehem, PA 18015

Contents

Foreword..iii

Plenary Paper

Bubble Dynamics in Liquid-Solid Suspensions
Liang-Shih Fan 1

Research and Development Papers

Infinitesimal and Finite Voidage Perturbations in the Compressible-Particle-Phase
Description of a Fluidized Bed
P. U. Foscolo, L.G. Gibilaro and S. Rapagna 44

Investigation of Dispersion Characteristics of a Fluidized Bed Reactor
C. Ercan, S.C. Arnold and H.E. Barner 51

Basic Approach on the Control Systems for Granulation Processes:
Case of Fluidized Bed Granulators
H.O. Kono and J.J. Su 61

A Two-Stage Fluidized Bed Process for Converting Hydrogen Chloride to Chlorine
M. Mortensen, H.Y. Pan, R.G. Minet, T.T. Tsotsis, Sidney Benson and Juan Llibre ...71

The Influence of Hydrodynamics on the Performance of an Interconnected Fluidized Bed
System for Regenerative Desulfurization in Coal Conversion Processes
O.C. Snip, R. Korbee, J.C. Schouten and C.M. van den Bleek 82

Measurement of Bottom Bed and Transport Disengagement Heights in Beds of Fresh FCC Catalyst
D. Geldart, Yan Xue and H.-Y. Xie 93

Dimension Measurements of Hydrodynamic Attractors in Circulating Fluidized Beds
Lu Huilin, D. Gidaspow and J.X. Bouillard 103

Gravity Flow of a Fluid-Particle Mixture in a Channel
R. Gudhe, K.R. Rajagopal, M. Massoudi and R.C. Yalamanchili 112

Microreactor Simulating Reaction Scene in Turbulent Fluid Bed of Group A Powder:
1. Axial Gas Dispersion
G. Gregory Benge and Arthur M. Squires 119

Microreactor Simulating Reaction Scene in Turbulent Fluid Bed of Group A Powder:
2. Usefulness in Fluid-Bed Reaction Engineering
Arthur M. Squires and G. Gregory Benge 128

Flow Rate Monitoring and Measurement in Dilute Phase Pneumatic Conveying
Using Pressure Fluctuations
Steve Tallon and Clive E. Davis 137

Simultaneous *In-Situ* Determination of Particle Loadings and Velocities in a Gaseous Medium
Ray Cocco, John Cleveland, Rich Harner and Ray Chrisman 147

A Light/Charge Solids Flow Meter
James Radar, George E. Klinzing and Arnand Prakash 154

Non-Intrusive Measurement of Solids/Liquid Concentration and Velocity:
Use of MRI Visualization
J.X. Bouillard and S. Altobelli 164

Characterization of Fluidization Properties of Fine Powder FCC Catalyst at Elevated Temperatures:
Prediction of Rheological Parameters
H.O. Kono, Y. Itani, E. Aksoy, E. Koresawa and J.J. Su 170

Index 180

Plenary Paper

Bubble Dynamics in Liquid-Solid Suspensions

Liang-Shih Fan

Department of Chemical Engineering, The Ohio State University, Columbus, OH 43210

Systems involving bubbles in liquid-solid suspensions frequently occur in a variety of multiphase operations in industry involving physical, chemical, petrochemical, electrochemical and biochemical processing. In both reactive and non-reactive systems, gas bubbles play an important role in determining the behavior or performance of the system. For example, gas bubbles are usually a source of reactant species whose transport across the interface often depends on the fluid flow around the bubble; gas bubbles induce intimate liquid/solids mixing; and in a three-phase fluidized bed, gas bubbles are responsible for bed contraction and solids entrainment to the freeboard. It has been recognized that the bubble wake located immediately underneath the bubble base is the dominating factor contributing to the system performance. It is, thus, of paramount importance to thoroughly understand the fluid dynamic behavior of the bubble and its associated wake.

This paper reviews the recent progress on the fluid dynamics of bubbles and bubble wakes in liquid-solid suspensions. The work of the author is discussed in the context of other researchers' work reported in the literature. The discussion presented covers both single bubble and multi-bubble systems; specifically, the subjects pertaining to single bubble dynamics include bubble rise characteristics and bubble breakup, wake pressure variation, particle entrainment, and heat transfer. Numerical simulations of bubble breakup in liquids along with an experimental verification of the computational results are also presented. A Particle Image Velocimetry (PIV) technique, instantaneous flow structures, and flow instability are discussed for multi-bubble dynamics. In addition, the bubble dynamics under high pressure (~ 200 atm) and high temperature (~180 °C) conditions elucidated based on flow visualization are presented.

INTRODUCTION

Bubbles in liquid-solid suspension represent a significant mode of three-phase contact characterizing a variety of physical, chemical, petrochemical and biochemical process operations (*e.g.*, Shah, 1979; Ramachandran and Chaudhari, 1983; Deckwer, 1985; Fan, 1989). These operations are conducted in three-phase reactors, such as a slurry bubble column or a three-phase fluidized bed. The more recent commercial developments using these reactors include the production of wax or diesel using the SASOL Slurry Bed Process (slurry bubble column in SASOL I) for Fischer-Tropsch Synthesis (Inga, 1994), the production of diesel or light hydrocarbon using the Texaco T-Star process (three-phase fluidized bed) for vacuum gas oil hydrotreating/hydrocracking (Johns *et al.*, 1993), and the production of human viral vaccines using mammalian/animal cell cultures in a three-phase bioreactor (Kologerakis and Behie, 1995).

In these systems, bubble dynamics play a key role in dictating the transport phenomena and ultimately affect the overall rates of reactions. It has been recognized that the bubble wake, when it is present, is the dominating factor governing the system hydrodynamics. In general, consideration of the flow associated with the bubble wake near the bubble rear, whether laminar or turbulent, is essential to characterize the complete behavior of the rising bubble including its motion. Conversely, examining the shape, rise velocity and motion of a bubble can provide an indirect understanding of the dynamics of the liquid-solid flow around the bubble. Temperature and pressure are the key operating variables dictating the physical properties of the gas and liquid phases and hence the bubble behavior. The liquid-solid flow medium through which the gas bubbles rise may be characterized as a pseudo-homogeneous medium under certain operating conditions. The most marked effects are flow instability induced in the downstream of the bubbles, and the resulting large-scale vortical motion and local solids concentration gradient. Interactions between neighboring bubbles are usually recognized in terms of bubble coalescence and/or breakup. Direct causes of these phenomena, however, lie in the interactions between the bubbles and the surrounding flow. The macroscopic flow structure under free bubbling conditions is a result of complex bubble-bubble and bubble-particle interactions. The instantaneous as well as time averaged flow fields, which vary greatly with gas velocity, demonstrate coherent structures under a range of gas velocities.

This paper is intended to review several fundamental aspects of the dynamics of single and multi-bubbles in liquid-solid suspensions which are of paramount importance to the transport processes in the system. Specifically, five subjects are included: they are (1) bubble rise characteristics, (2) bubble induced liquid-solid flow and transport phenomena, (3) bubble breakup mechanism due to bubble-particle collision, (4) instantaneous flow phenomena, and (5) pressure and temperature effects. The focus of each subject may be different, but each subject is phenomenologically interrelated. The review is presented in the context of the recent work contributed by the author's research group and is discussed in light of other researchers' work reported in the literature on these subjects.

BUBBLE RISE CHARACTERISTICS

The characteristics of a rising bubble can be described in terms of the shape, rise velocity and motion of the bubble. These rise characteristics are closely associated with the flow and physical properties (mainly viscosity or presence/absence of solid particles) of the surrounding medium as well as the interfacial properties (i.e., presence/absence of surfactant) of the bubble surface.

Bubble Shape

The interaction between a rising gas bubble and its surrounding liquid-solid medium determines the shape of the bubble and the extent of the disturbance in the surrounding flow field. It is probably most convenient to describe the behavior of a bubble and its wake in light of its shape.

Bubbles in motion in Newtonian liquids are generally classified by shape as spherical, oblate ellipsoidal, and spherical/ellipsoidal cap. The observed bubble shape is a result of an intricate balance among forces acting on the bubble, viz., surface tension, viscous and inertial (or buoyancy) forces. Figure 2.1 shows sketches of bubble shapes thus classified. As the dominant forces change with increasing bubble size, the bubble shape goes through a transition from spherical to ellipsoidal and then to a spherical-cap shape.

When the bubble size is small (for example, d_e less than 1 mm in water), surface tension forces predominate and the bubble shape is approximately spherical. The flow around the bubble is controlled by viscous forces and can be described, provided Re_e ($=\rho_l d_e U_b/\mu_l$) $\ll 1$, by the creeping flow approximation, i.e., by the Hadamard-Rybczynski theory for purified systems or by Stokes' law for contaminated systems (Clift et al., 1978). For small spherical bubbles at very low Reynolds numbers, therefore, the flow does not separate from the bubble surface and the wake does not exist.

For bubbles of intermediate size, the effects of both surface tension and the inertia of the medium flowing around the bubble are important. In addition, the bubble dynamics are primarily influenced by the liquid viscosity and the presence of surface-active contaminants. As a result, intermediate-size bubbles exhibit very complex shape and motion characteristics. Though called ellipsoidal bubbles, they often lack in fore-and-aft symmetry and, in extreme circumstances, cannot be described by any simple regular geometry due to significant shape fluctuations. Correspondingly, the behavior of wakes behind intermediate-size bubbles is complicated.

Large bubbles, whose volume is greater than 3 cm³ (i.e., $d_e > 18$ mm) in general (Clift et al., 1978), are governed by inertial or buoyancy forces with negligible effects of surface tension, viscosity and purity of the liquid media. The bubble shape is approximately a spherical cap with an included angle of about 100° (provided $\mu_l < 50$ mPa·s) and a relatively flat or sometimes indented base. These spherical-cap bubbles generate relatively large-volume wakes.

In systems other than gas-liquid systems, such as gas-slurry bubble columns (Dayan and Zalmanovich, 1982) and gas-liquid-solid fluidized beds of small, light particles (Stewart and Davidson, 1964; Ostergaard, 1973), bubble behavior has often been observed to resemble that in viscous liquids, e.g., the large bubbles assume the spherical-cap shape mentioned above. This similarity is based on the premise that the liquid-solid mixture in such systems can be regarded as a pseudo-homogeneous medium of apparently higher viscosity compared to the suspending liquid. On the other hand, in three-phase fluidized beds of large and/or heavy particles, gas bubbles behave in a different way. In a bed of dense particles such as lead or iron shot, Stewart and Davidson (1964) observed that a liquid bubble could be stabilized by either small gas bubbles or a thin layer of gas lining its roof.

Although the flattening of the bubble with increasing bubble size can be explained qualitatively (Fan and Tsuchiya, 1990), quantitative predictions are not possible, especially when a variety of physical properties of the surrounding medium are involved. In this regard, the state-of-the-art knowledge about bubble shapes in various liquids is established from experimental observations. Three dimensionless groups, viz., the Reynolds number Re_e, the Eötvös number ($Eo = g\rho_l d_e^2/\sigma$), and the Morton number ($Mo = g\mu_l^4/\rho_l\sigma^3$) (Grace, 1973; Bhaga and Weber, 1981) are commonly used to characterize the bubble rise behavior.

For bubbles in low Mo liquids ($Mo < 10^{-3}$), Vakrushev-Efremov's (1970) formula given by

$h/b =$

$$\begin{cases} 1 & Ta < 1 \\ \{0.81 + 0.206\tanh[2(0.8 - \log_{10} Ta)]\}^3 & 1 \le Ta \le 39.8 \\ 0.24 & 39.8 \le Ta \end{cases}$$

(2.1)

can be used to predict the bubble aspect ratio. Here, Ta ($= Re_e Mo^{0.23}$) is the Tadaki number.

The bubble characteristics in two-dimensional (2-D) systems were studied by various researchers (e.g., Henriksen and Ostergaard, 1974; Tsuchiya and Fan, 1988; Song, 1989; Tsuchiya et al., 1990). The recent results of Tsuchiya et al. (1990) for glass beads of four different sizes, each at three different bed voidages ($\epsilon_l = 1 - \epsilon_s$), in a 2-D (tap) water-solid fluidized bed indicate that, for a given particle size, the variation in bed voidage has a minimal influence on the bubble aspect ratio over the range $0.5 < \epsilon_l \le 1$. The effect of particle size is also not significant for most particles that Tsuchiya et al. (1990) examined; however, at higher Re_e (> 3000) the bubbles flatten less extensively for the larger particles (1 mm glass beads with terminal velocity of 0.17 m/s).

The bubble shape in three-dimensional (3-D) fluidized beds with higher solids holdups was investigated by Zhang et al. (1994) for bubbles just emerging from the bed surface. Figure 2.2 shows such results for water-1 mm glass bead fluidized bed with different solids holdups. Their results indicate that the bubbles in the fluidized bed are flatter than those in single fluids when the bubble Tadaki numbers are less than 10. For bubbles with Tadaki numbers larger than 10, the bubbles are less flatter than those in single fluids. Such a deviation of the bubble aspect ratio increases with an increase in the particle size and the solids holdup.

The effect of particle wettability was investigated by Tsutsumi et al. (1991) using 460 and 774 μm glass beads (wettable) and Teflon-coated 460 and 774 μm glass beads (non-wettable). They found no appreciable difference in the bubble aspect ratio, and the same qualitative dependence on the particle size and the solids holdup was observed between these two types of liquid-solid systems.

Bubble Rise Velocity

Estimating the bubble rise velocity is essential; however, the theoretical development of a universal prediction equation over a wide range of bubble size or Reynolds number has been hampered due to the diversified effects of the wake flow. At low or high Re_e, the bubble rise velocity can be determined from the fore-and-aft symmetry of the flow around spherical bubbles without considering the wake effect, or based on the assumption of potential flow around the nose of spherical-cap bubbles. At intermediate Re_e, the presence of the bubble wake causes an increase in the form drag and has a significant effect on the rise velocity. Therefore, it is practical to develop empirical correlations which feature the dominant force(s) acting on the bubble over wide ranges of bubble sizes and physical properties of the surrounding medium. Fan and Tsuchiya (1990) proposed a semi-empirical correlation which has the form

$$U_b = \left[\left(\frac{\rho_l g d_e^2}{K_b \mu_l}\right)^{(-n)} + \left(\frac{2c\sigma}{\rho_l d_e} + \frac{g d_e}{2}\right)^{(-n/2)}\right]^{(-1/n)} \quad (2.2)$$

Three parameters n, K_b, and c depend on physical properties of the liquid and/or contamination level of the liquid.

A liquid-solid mixture of small, light particles can often be treated as a pseudo-homogeneous medium. The behavior of a rising bubble in a "homogeneous" or "heterogeneous" medium depends primarily on the ratio of the particle diameter to bubble diameter (d_p/d_e). In principle, the homogeneity assumption of the fluidized systems can be verified by comparing the single bubble rise characteristics in the media with those in infinite liquids. However, this approach can only be applied under a limited range of conditions because of the lack of correlation(s) for the bubble rise velocity in various liquids over a wide range of bubble sizes. Comparing the rise velocities of spherical-cap bubbles in a fluidized bed with those in viscous liquids, Darton and Harrison (1974) calculated the effective bed viscosity for liquid-solid systems. With the apparent viscosity of the liquid-solid fluidized bed, the bubble rise characteristics and the associated bed rheology can be qualitatively studied using the graphical correlation developed for bubbles rising in infinite liquids (Grace, 1973; Bhaga and Weber, 1981). Figure 2.3 shows the characteristics of a single bubble rising in infinite Newtonian liquids in comparison with those in a 3-D liquid-solid fluidized bed over a wide range of solids holdups, particle sizes and particle densities. For each representative bubble shape (denoted by its acronym), refer to the corresponding sketch presented in Fig. 2.1. To characterize the liquid-solid interfacial properties, the surface tension of water is used since the particles are wettable. The apparent bed density (ρ_m) is defined as

$$\rho_m = \rho_s \epsilon_s + \rho_l \epsilon_l \quad (2.3)$$

and the apparent bed viscosity (μ_b) is evaluated as

$$\mu_b = \mu_l \exp(36.15 \epsilon_s^{2.5}) \quad (2.4)$$

Equation (2.4) is an empirical correlation proposed by Darton (1985) and is applicable for particles of moderate densities ($\rho_s < 3$ g/cm^3) and sizes ranging from 50 to 1000 μm, and for solids holdups higher than 0.2.

Figure 2.3 reflects that a liquid-solid suspension can be regarded as a pseudo-homogeneous phase only under limited conditions. In a water-glass bead (550 μm) fluidized bed with a high solids holdup (ϵ_s=0.526), Darton and Harrison (1974) reported that the rise velocity of small bubbles is much smaller than that in an equivalent Newtonian medium. Darton (1985) attributed this effect to the non-Newtonian behavior of the fluidized beds near the point of incipient fluidization. More generally, from Fig. 2.3, it can be seen that for $Eo_m < 10$, the variation in the trend of the data for liquid-solid suspensions with Eo_m is not parallel to that exhibited by the solid curves for Newtonian liquids. For $Eo_m < 10$, the bubble Reynolds number in the fluidized beds in some cases is smaller than that in a single Newtonian liquid, which indicates a higher apparent bed viscosity for a bed of small bubbles. Thus, for these cases the fluidized bed exhibits the properties of shear-thinning fluids with respect to the rising bubbles. On the other hand, in cases for $Eo_m < 10$ when the bed is composed of small particles and has a lower solids holdup (e.g., $d_p = 163$ μm at $\epsilon_s = 0.42$, in Fig. 2.3), the bubble tends to find its way through interstices between the particles and rises faster as it would rise through a "fictitious" homogeneous medium. Under this situation, the fluidized bed exhibits shear-thickening properties with respect to the rising bubbles.

Zhang et al. (1994) correlated the values of K_b in Eq. (2.2) in terms of the particle terminal Reynolds number and the solids holdup for water-glass bead fluidized systems using the apparent bed density and viscosity defined by Eqs. (2.3) and (2.4), respectively. For most high viscosity liquids, the parameters n and c are recommended to be 1.6 and 1.4, respectively (Fan and Tsuchiya, 1990). These values are chosen for the water-glass bead fluidized bed with a solids holdup higher than 0.4. The correlation for K_b is

$$K_b = 0.0033 \, Re_t^{0.3} \, \epsilon_s^{-9.6} \quad (2.5)$$

When the solids holdup of the water-glass bead fluidized bed is lower than 0.4, the bubble rise behavior is very similar to that in low viscosity liquids. For most low viscosity liquids, 1 and 1.04 are recommended as values of n and c, respectively. The K_b in Eq. (2.2) for water-glass bead fluidized beds at solids holdups higher than 0.07 and less than 0.4 is correlated as

$$K_b = 3.66 \, Re_t^{0.51} \, \epsilon_s^{-0.54} \quad (2.6)$$

Figure 2.4 shows that the predictions from the present correlations agree well with the experimental results for single bubbles rising in water-glass bead fluidized beds for various solids holdups.

The effect of particle wettability on the bubble rise velocity was examined in a 2-D system by Tsutsumi et al. (1991). Non-wettable particles have a large contact angle compared to wettable particles. Thus, non-wettable particles favor the contact between bubbles and solids. Tsutsumi et al. (1991) observed that in small bubble systems, non-wettable particles usually attached to the bubble and form a monolayer on the bubble surface. Consequently, the bubble rise velocity is smaller than that for wettable particles. On the other hand, negligible effect of particle wettability was observed on the rise velocity of large bubbles ($d_e > 15$ mm). For large bubbles with circular-cap shape, the attachment of particles to the bubbles occurred only at the bubble base, and was not observed on the bubble roof due to the fluid shear effects caused by the fast rising bubbles.

BUBBLE BREAKUP MECHANISM DUE TO BUBBLE-PARTICLE COLLISIONS

In a three-phase fluidization system, bubble size variation, which strongly influences the flow regime transition and mass and heat transfer phenomena between phases, is intimately related to bubble-particle collisions. The collisions can yield two different consequences: the particle is ejected from the bubble surface, or the particle penetrates the bubble leading to either bubble breakage or non-breakage.

Bubble-particle collisions generate perturbations on the bubble surface. After the bubble-particle collision, three factors become crucial to determine the breakage characteristics of the bubble (Clift *et al.*, 1978):

1. Shear stress which depends on the liquid velocity gradient tends to break the bubble. The magnitude of the stress increases as the relative impact speed increases. Therefore, under the conditions discussed in the present paper, higher particle density and larger bubble size will result in a higher impact speed and, thus, a higher shear stress.

2. Surface tension force tends to stabilize the bubble and recover the original shape.

3. Viscous force slows the growth rate of the surface perturbation, and tends to stabilize the bubble.

Chen and Fan (1989) established criteria for particle penetration depending on three different conditions. The criteria were developed neglecting the shear effects due to liquid flow. When any of these three criteria is satisfied, the particle will penetrate the bubble. When none of these criteria is satisfied for a particle in contact with the bubble, the particle will be ejected from the bubble surface (see Figs. 3.1(a) and (b) for collision configuration).

(i) Acceleration of the particle is downward

$$a_0 = g\frac{\rho_l - \rho_s}{\rho_s} + \frac{6\sigma(1 + d_p/2R)}{\rho_s d_p^2} < 0 \qquad (3.1)$$

(ii) Particle velocity relative to the bubble is downward

$$\frac{(U_b + U_{p0})\omega}{a_0} > 1.0 \qquad (3.2)$$

where

$$\omega = \left(\frac{3}{2}\frac{\rho_l}{\rho_s}\frac{g}{d_p}\right)^{1/2} \qquad (3.3)$$

(iii) Particle penetration depth is larger than the deformed bubble height

$$\begin{aligned}h_p &= \frac{(U_b + U_{p0})}{\omega}\sinh(\omega\tau_p) \\ &\quad - \frac{a_0}{\omega^2}[\cosh(\omega\tau_p) - 1] \\ &= \frac{a_0 - [a_0^2 - (U_b + U_{p0})^2\omega^2]^{1/2}}{\omega^2} > H\end{aligned} \qquad (3.4)$$

By extending Boys' instability analysis (Boys, 1959), Chen and Fan (1989) obtained a criterion for bubble breakage after the penetration as [see Fig. 3.1(c)]

$$d_p > H_d \qquad (3.5)$$

Hong and Fan (1994) conducted experimental and numerical studies to elucidate the effects of flow field and liquid viscosity on the bubble breakage. The system in their study included a freely rising bubble colliding with a falling copper ball of density 8.7 g/cm^3. Particles of three different sizes (0.92, 1.22, and 1.53 cm in diameter) were used, with the spherical-cap bubble volume equivalent diameter fixed at 1.97 cm. Two liquids 80% (wt) aqueous glycerine solution and pure water were used. The surface tensions of these two liquids are comparable ($\sigma_{solution}$=65.9 mN/m vs. σ_{water}=72.6 mN/m); however, the viscosities differ greatly ($\mu_{solution}$=52.9 mPa·s vs. μ_{water}=1.0 mPa·s).

Figure 3.2 shows the experimental results in the glycerine solution. Figures 3.2(a) and (b) are for particle sizes of 1.22 cm and 1.53 cm, respectively. It is seen from the figures that, after the collision, the particle penetrates the bubble leaving a hole in the bubble. During the penetration, the bubbles, in both figures, are locally stretched/elongated by the downward motion of the liquid induced by the particle due to the viscous effects. However, the results after the penetration are different. In Fig. 3.2(a), the funnel-shaped bubble recovers its original shape without breakage; however, in Fig. 3.2(b), the bubble is broken down into pieces. In these experiments, the particle Re numbers are less than 500 yielding a laminar liquid flow around the particle. The viscous force is dominating and the perturbations generated by the collision on the gas-liquid interface are dampened by the viscous effect. After the penetration, the bubble stability is dictated by Boys' instability in the horizontal direction. Equations (3.1) through (3.5) describe these phenomena well.

The bubble in the glycerine solution can also be broken by other mechanisms as shown in Fig. 3.3. In this figure, the particle collides with a side portion of the bubble instead of the center. The bubble is elongated in the vertical direction. The breakage of such an elongated bubble is due to Boys' instability in the vertical direction.

A comparison between Fig. 3.2(a) and Fig. 3.4 reveals the different bubble breakage mechanism in water. Although the only significant difference in conditions between these two cases is the liquid viscosity, the consequences of the particle penetration are opposite: in Fig. 3.2(a), the bubble recovers to its original shape; while in Fig. 3.4, the bubble is broken down into pieces. In Fig. 3.4 the particle Re number is 1.2×10^4 and the liquid flow around the particle is turbulent, while in Fig. 3.2(a) the particle Re number is less than 500 and the flow is laminar. Hence, the bubble in water shown in Fig. 3.4 is disintegrated by the surface perturbations induced by the turbulent shear stresses when the particle penetrates. The bubble breakage due to perturbation induced by the turbulent shear stress can be further illustrated by comparing Fig. 3.3 and Fig. 3.5, when the particle collides with a side portion of the bubble in water. It is seen that in Fig. 3.5, the bubble is less elongated in vertical direction than that in Fig. 3.3.

Numerical Simulation

Since the characteristic time of a typical bubble-particle collision is less than 0.1 second and the particle and bubble sizes are small (on the order of 1 cm), it is extremely difficult to obtain the detailed information on the collision process experimentally. In order to obtain a better understanding of this process, numerical simulation of the bubble-particle collision process is conducted.

In this study, the VOF (Volume of Fluid) method (Hirt and Nichols, 1981) is employed to simulate the dynamic behavior of bubbles. Using the local instantaneous Navier-Stokes equations and the cell-averaged volumetric fractions of each phase, the motion of the gas-liquid interface is traced out directly. The governing mass and momentum balance equations are

$$\frac{\partial \rho_l}{\partial t} + \nabla \cdot (\rho_l U_l) = 0 \qquad (3.6)$$

$$\frac{\partial}{\partial t}(\rho_l U_l) + \nabla \cdot (\rho_l U_l U_l) = \\ -\nabla p + \nabla \cdot [\mu_l(\nabla U_l + \nabla U_l^T)] + \rho_l g \qquad (3.7)$$

The basic solution algorithm is based on 3-D SOLA method. The bubbles are treated as void regions filled with ideal gas and the gas motion inside the bubbles is neglected. The progressive deformation of a bubble is assumed to be isentropic.

The movement of the gas-liquid interface is tracked based on the distribution of F, the volume fraction of liquid in a computational cell. Therefore, the bubble surface is represented by the cells where F lies between 0 and 1. From the distribution of F, the size and shape of the bubble can be determined. The curvature of interface is then used to calculate the surface tension force. The effect of the surface tension force is introduced into the computation of velocity and pressure fields by an equivalent surface pressure. The governing equation of F is given by

$$\frac{\partial F}{\partial t} + \nabla \cdot (FU_l) = 0 \qquad (3.8)$$

With the proper selection of a reference coordinate system, configurations of computational domain, and boundary conditions, the computations are conducted on CRAY Y-MP8/864 at the Ohio Supercomputer Center using the FLOW-3D codes (Harper et al., 1991). The simulation is conducted in the Cartesian coordinate system. By symmetric approximation, only one fourth of the cylindrical column is simulated, the number of grids are taken to be $12 \times 12 \times 52$. For simplicity, it is assumed that the particle is moving downwards at a constant speed (1.0 m/s) throughout the collision. This means that the motion of the particle is not influenced by the presence of the bubble, which is valid only when the particle density is sufficiently high. In the simulation, the particle is fixed in the computational domain and the motion of the particle is accounted for by superimposing an equivalent constant upward velocity on the liquid phase. In the simulation, water is used as the liquid phase, and the bubble Reynolds number is on the order of 1×10^4, which indicates that the simulation is performed under low viscosity conditions.

Figure 3.6 shows the numerical simulation results of the bubble-particle collision process with bubble breakage ($d_e/d_p = 1.81$). The velocity vector and pressure fields during the collision process are presented in the figure. The simulation reveals that, as the particle comes in contact with the bubble, the bubble starts to deform into a bowl shape. The bubble is then penetrated by the particle and is further deformed into a ring shape. The ring-shaped bubble is exposed to a strong shear flow induced by a downward liquid flow inside the wake region of the colliding particle and an upward flow outside the wake region. The strength of the wake flow depends on the relative contact velocity of the particle to bubble, particle size, and liquid viscosity. The surface tension force also plays an important role in the deformation of the ring-shaped bubble. Under this simulation condition, it is revealed that the surface tension force and viscous force are unable to overcome the shear stress induced by the wake flow, and eventually the bubble is broken.

In Fig. 3.7, the simulation of a small particle colliding with a large bubble ($d_e/d_p = 3.33$) is shown. The bubble size is similar to that in Fig. 3.6, but the particle size is much smaller. Because of the small size of the particle and the particle wake, the shear stresses on the ring-shaped bubble are relatively small. As a result, the surface tension force and the viscous force are able to overcome the shear stresses induced by the wake flow and the bubble ring merges back into its original shape.

BUBBLE INDUCED LIQUID-SOLID FLOW AND TRANSPORT PHENOMENA

A major factor contributing to the complexity of the liquid-solid flow in the vicinity of the rising bubbles is the instability induced by the wakes behind the bubbles. Through this instability, the wake flow becomes unsteady and interactive with the external flow. The phenomenon is often characterized by a cycle of vortex formation and shedding and, for free-rising bubbles, it is intimately related to the oscillatory bubble motion. Various transport phenomena taking place in gas-liquid-solid systems are closely associated with the wake flow behavior. Detailed accounts of these phenomena as well as wake flow behavior are given in the book by Fan and Tsuchiya (1990). In the following, two recent studies concerning the pressure distribution around a bubble in liquid-solid suspension and the wake enhanced heat transfer phenomena are presented. The drift effects associated with rising bubbles are also discussed.

Wake Pressure

Pressure distribution in the wake is critical and is closely associated with the fluid motion, solids concentration in the bulk phase and the size of the wake, or more generally, the wake structure. Lazarek and Littman (1974) studied the pressure field around a large circular-cap bubble in water and verified the Davies and Taylor boundary condition along the bubble roof. Bessler and Littman (1987) later extended this study for liquids of different viscosities. The pressure signal showed a symmetric minimum in the primary wake. They also observed a sharp minimum followed by localized recovery of the pressure immediately beneath the bubble base, but only for low viscosities. In their investigations, the bubble wake, even for water, was closed laminar in structure, in spite of the large bubble size (up to 10 cm wide). The closed laminar wake was possibly caused by the delayed onset of vortex shedding from the wall stabilized large bubble.

Raghunathan *et al.* (1992) reported the pressure distribution around a bubble in water-glass bead (163 μm) suspension. Figure 4.1 shows a typical pressure distribution, in which the time and pressure values are reduced to dimensionless forms τ and pressure coefficient C_p, respectively, as

$$\tau = \frac{U_b t}{a} \tag{4.1}$$

$$C_p = \frac{p - p_\infty}{\frac{1}{2}\rho_m U_b^2} \tag{4.2}$$

In the above equations a is the bubble half width and $(p-p_\infty)$ is the differential pressure; ρ_m and U_b are the liquid-solid mixture density and relative bubble rise velocity, respectively. For liquid-solid systems, as proposed by El-Temtamy and Epstein (1980),

$$\rho_m = \rho_s \epsilon_s + \rho_l (1 - \epsilon_s) \tag{4.3}$$

$$U_b = U_b' - \frac{U_l}{1 - \epsilon_s} \tag{4.4}$$

where U_b' is the observed rise velocity of the bubble and U_l is the liquid velocity. Note that ρ_m and U_b reduce to ρ_l and U_b' respectively for the case of stationary water.

It is seen in the figure that while the first maximum (τ_o, C_{po}) in the pressure signal lies in the frontal field of the bubble, the first minimum (τ_1, C_{p1}) is located just beneath the bubble base within the stable liquid layer region defined by Tsuchiya and Fan (1988). The second maximum (τ_2, C_{p2}) is at the boundary between the stable liquid layer and the vortex and thus corresponds to the beginning of the circulatory region. The second minimum (τ_3, C_{p3}) is located in the horizontal plane in which the vortex center lies; however, this concave region is not as sharp as the first minimum. Tsuchiya and Fan (1988) defined the boundary of the primary wake as the cut-off stream flowing across the wake and the point (τ_4, C_{p4}) in the pressure distribution is at the cut-off stream as shown in the photograph. Thus, the regions corresponding to $(\tau_2-\tau_1)$, $(\tau_4-\tau_2)$, and $(\tau_4-\tau_1)$ measure the sizes of the stable liquid layer, the vortex, and the primary wake respectively. This pressure distribution depicted in Fig. 4.1 is qualitatively quite similar to those reported by Lazarek and Littman (1974) in water. However, the pressure recovery is nearly complete in a liquid-solid suspension indicating a smaller secondary wake compared to that in water.

Drift Effect

The bubble wake behavior directly affects the particle elutriation in the operation of a three-phase fluidization system and may be significant if the freeboard region is not sufficiently large. The problem appears to be more significant for beds of small and/or light particles than those of large and/or heavy particles. Furthermore, the axial solids holdup distribution differs appreciably between the former and latter types of particles in the freeboard region.

Page and Harrison (1974) studied the fundamental mechanisms for particle entrainment and de-entrainment in the transitional region of the freeboard. They indicated that the particles were drawn from the upper surface of the fluidized bed into the freeboard in the wake behind the bubble and the vortices containing the particles were shed from the wake in the freeboard. They also found that particle entrainment decreased with a decrease in bubble size and bubble frequency and with an increase in liquid velocity and particle size. El-Temtamy and Epstein (1980) developed a model to predict the solids holdup distribution in the freeboard. In the model, they clearly identified the critical roles played by the bubble wake in particle entrainment and the wake shedding in particle de-entrainment.

Some physical insight into the mechanisms of particle carryover can be gained by closely following the time evolution of the particle flow around a single bubble. Miyahara *et al.* (1989) examined a series of photographs of such a time sequence of particle entrainment and de-entrainment by a single bubble into the freeboard of a 2-D water-774 μm glass bead fluidized bed. They reported that when the bubble emerges from the upper free surface of the fluidized bed, a mantle of particles covering the bubble roof drains away and rushes into the near wake of the bubble. Overall, the particles move upward due to this near-wake capture as well as due to the drift effect (Darwin, 1953). The latter, however, is confined to the vicinity of the bed surface. The drift effect can be distinctly seen when low density calcium alginate particles (1.2 mm in diameter and 1.02 g/cm^3 in density) are used, as shown in Fig. 4.2 (Tsuchiya et al., 1992). The particle displacement caused by the drift effect is thus relatively insignificant in case of large bubbles and/or heavy particles.

Bubble Wake Induced Heat Transfer

The heat transfer characteristics are closely associated with the fluid flow behavior. The following considers the interaction between a particle and the bubble wake and its effect on the local heat transfer in a liquid or liquid-solid suspension.

Comprehensive literature reviews on the heat transfer behavior in three-phase fluidization systems given by Kim and Laurent (1991), and more recently by Kumar et al. (1992, 1993a, 1993b) and Kumar and Fan (1994) report the numerous heat transfer studies in various gas-liquid and gas-liquid-solid systems. Most prior studies have dealt primarily with the steady-state time-averaged heat transfer behavior between the heat transfer surface and the bed. Values of heat transfer coefficient between the bed and the column wall have been reported by Ostergaard (1964), Viswanathan et al. (1964), Kato et al. (1981), Chiu and Ziegler (1983), and Muroyama et al. (1984, 1986); while those between the surface of an immersed heating object and the bed have been reported by Baker et al. (1978), Kato et al. (1984), Kang et al. (1985), and Magiliotou et al. (1988). Heat transfer in a three-phase fluidization system has a strong dependence on the physical properties of the liquid phase and a weak dependence on the gas phase properties. The heat transfer coefficient has been determined to increase with the gas and liquid velocities, the size and density of the particles, the diameter of the column, and the thermal conductivity and heat capacity of the liquid but decreases with an increase in liquid viscosity and the diameter of the immersed object (Kim and Laurent, 1991). Baker et al. (1978) reported that the heat transfer coefficient initially increased quite rapidly with gas velocity but became less marked at higher gas flow rates, asymptotically approaching a maximum value. They further reported that the plot of heat transfer coefficient vs. liquid velocity exhibited a maximum, similar to that observed in liquid-solid fluidized beds. The bed voidage at which the maximum heat transfer occurred decreased with increasing particle size, but increased with the liquid viscosity (Kang et al., 1985). Chiu and Ziegler (1983) reported higher heat transfer from wall to bed in a three-phase fluidized bed compared to that in liquid-solid and gas-liquid systems except under the condition of small particles and high gas velocities. They postulated that the product of the heat transfer coefficient and the liquid holdup remains the same in liquid-solid and gas-liquid-solid systems and proposed a model based on the thermal resistance of the bed.

Deckwer (1980) proposed a heat transfer mechanism in bubble columns based on Higbie's surface renewal theory combined with Kolmogoroff's theory of isotropic turbulence. Several other investigators (Suh et al., 1985; Magiliotou et al., 1988; Suh and Deckwer, 1989) extended this concept to three-phase fluidized beds by modifying the energy dissipation rate to include the increased surface renewal due to the increased turbulence created by the solid particles. Suh et al. (1985) and Suh and Deckwer (1989) neglected the effect of particle convective transport on the heat transfer, while Magiliotou et al. (1988) proposed that the particles also contribute to heat transfer in both liquid-solid and gas-liquid-solid fluidized beds in conjunction with isotropic fluid micro-eddies. Being semi-theoretical in approach, these studies do not fully account for the inherent mechanism underlying the heat transfer in three-phase fluidized beds. Furthermore, these previous studies usually employed large heat transfer probes to measure the steady-state heat transfer coefficient, thereby, losing the information pertaining to the time-variant local hydrodynamic effects (e.g., bubble motion, bubble size, and wake type) on the heat transfer coefficient.

Kumar et al. (1993a, 1993b) and Kumar and Fan (1994) asserted that the heat transfer property in the bed is intimately associated with the bubble motion, bubble size, and phase holdups, which are affected by the hydrodynamic behavior of the system, including wake flow. They quantified the bubble wake effects on the heat transfer by the measurement of the instantaneous local variations in the heat transfer coefficient due to the passage of gas bubbles in liquid and liquid-solid systems. A micro-foil heat transfer probe was developed to trace the instantaneous local heat transfer rate during the passage of gas bubbles. Simultaneous visualization was performed to establish the correspondence between the visual and probe signals, and hence relate the local instantaneous hydrodynamics to the heat transfer rate. Figure 4.3 shows a representative example of the time dependent heat transfer coefficient in a liquid-solid fluidized bed of low density gel beads, with associated photographs of the bubble-wake induced liquid-solid flow pattern in the vicinity of probe corresponding to the heat transfer coefficient obtained. In the photographs the bright dots are the gel particles and the bright vertical object is the heat transfer probe located at the center of the column. The injected bubble volume is 3 cm^3. Photograph A shows the instantaneous flow field where a marked increase in the heat transfer coefficient results as the bubble approaches the heat transfer surface. It is also observed that the instantaneous heat transfer coefficient starts increasing well before the bubble approaches the lower edge of the probe, although the photograph for this case is not shown here. This increase in heat transfer coefficient is attributed to the local turbulence caused by the approaching bubble. The bubble is spherical cap and the wake structure appears to be symmetric about the vertical axis of bubble movement. Photograph B shows a large vortex in the primary wake. The vortex entrains liquid and solids from around the wake causing rapid surface renewal at the probe surface which results in increased heat transfer. In the two-dimensional plane of visualization the probe surface appears to lie in the middle of the vortex structure close to the wake central axis. Photograph C shows the flow pattern near the maximum in the heat transfer signal. The heat transfer surface experiences high shear flow due to the primary wake induced upward liquid and solid flow toward the wake central axis causing enhancement in the heat transfer. The velocity of gel particles near the heat transfer surface is roughly estimated from the streak of particles. The upward particle velocity is in the range of 20 cm/s which is close to the bubble rise velocity (24 cm/s) confirming that the particles are in the wake central region. Thus, the maximum heat transfer is obtained along the wake central axis in the primary wake region.

The mechanistic model developed by Wasan and Ahluwalia (1969) based on the consecutive film and surface-renewal theory can be used to account for the heat transfer behavior in bubble columns and three-phase fluidized beds. In the application of this model, the heat transfer surface-fluid element contact time can be approximated by the ratio of heat transfer surface length to bubble rise velocity. The heat transfer film thickness can be approximated by the thermal boundary layer.

INSTANTANEOUS FLOW PHENOMENA

Three different flow regimes, namely, dispersed bubble, coalesced bubble and slugging are commonly identified for slurry bubble columns or three-phase fluidized beds. A gross circulation flow pattern was observed for these systems under both the dispersed bubble and the coalesced bubble regimes (*e.g.*, Latif and Richardson, 1972; Hills, 1974; Rietema, 1982; Reese *et al.*, 1993). Generally, the gross circulation flow field comprises an upward flow in the column core and a downward stream along the wall with an inversion point (zero liquid velocity) located at about 0.5 to 0.7 radius of the column (Walter and Blanch, 1983). The phase mixing, heat transfer and mass transfer characteristics are significantly influenced by the circulation pattern. The liquid circulation pattern is affected by the design and the operating variables such as column diameter, distributor, superficial velocities and physical properties of the phases. The dynamic flow pattern and the manner in which these variables affect the circulation, however, are not fully understood.

Early experimental work on quantitative measurements of the axial velocity distribution can be represented by that of Hills (1974). These data, obtained from time-averaged, point measurement conditions using a modified pitot tube (Pavlov tube), have been widely adopted as a data source for model verification. Through point measurements using hot-film anemometry, the "instantaneous" axial, tangential and radial velocities and turbulent intensity profiles in a bubble column were obtained by Franz *et al.* (1984). They concluded that the "instantaneous" liquid velocity profiles were generally asymmetrical and dynamic in nature. The axial turbulent intensity was significantly higher than those in the radial and tangential directions. The radial and tangential turbulent intensities were rather uniform in the radial direction; however, the axial intensity varied significantly. The work is important as it reveals the dynamic nature of the flow structure.

In contrast to the point measurements, Ulbrecht *et al.* (1985) studied the flow structure through direct flow visualization in a 3-D bubble column. Using viscous and non-Newtonian liquids, they identified three instantaneous flow patterns i.e., viscous pattern, helical flow pattern and vortex pattern in the column. They concluded that the range of existence of these three flow patterns depends on the gas flow rate, the diameter of the column and the viscosity of the liquid. Tzeng et al. (1993), through flow visualization, studied the macroscopic flow structures of 2-D gas-liquid and gas-liquid-solid fluidization systems. They found that the gross circulation pattern, which occurs at high gas velocities, is closely associated with the induced liquid or liquid-solid flows resulting from rising bubbles and bubble wakes. Based on the bubble dynamics and local liquid flow patterns, four distinct flow regions were identified when the gross circulation occurs, i.e., central plume region, fast bubble flow region, vortical flow region, and descending flow region.

To provide a quantitative measure of the flow field in a three-phase fluidized bed, non-intrusive techniques have been used to provide the time/volume averaged flow information. Latif and Richardson (1972) applied the refractive index matching technique and special optical arrangement to measure the 3-D flow field of the solid phase in a 3-D liquid fluidized bed for solid holdups up to 45% by volume. A γ-radiated particle was used as the tracer particle. The 3-D flow information was obtained from the time/volume averaging processes. They demonstrated that the solid phase moves in a single-cell gross circulation flow pattern with negligible radial circulatory motion in the lower part of the bed. In the upper part of the bed, however, no well-marked flow pattern was observed.

Recently, Devanathan et al. (1990) adopted a radioactive particle tracking (CARPT) technique originally developed by Lin et al. (1985) to study the liquid motion in bubble columns. In the CARPT technique a single particle is tracked over a long period of time to obtain the time/volume averaged liquid flow fields in a bubble column. Devanathan et al. (1990) reported the existence of a pair of circulation cells with the liquid descending along the column wall and ascending in the central region at gas velocities greater than 0.05 m/s. At lower gas velocities, however, they reported that two pairs of cells are present; the lower cell pair appears in the entrance region with liquid ascending along the wall and descending in the center. Yang et al. (1993) extended the study of bubble columns using the CARPT technique by investigating the Reynolds stresses and fluctuating velocities determined from the time averaged data. Using a similar technique, Larachi et al. (1994) measured the time/volume averaged solids flow patterns in a gas-liquid-solid fluidization system. These averaging techniques have provided useful information on the macroscopic behavior of gas-liquid and gas-liquid-solid fluidization systems but it has become apparent that better instrumentation is required to quantify the transient multiphase flow phenomena which is vital to the further development of fundamental theories (Tarmy and Coulaloglou, 1992). The Particle Image Velocimetry (PIV) technique has the ability to provide instantaneous, full flow field information.

Particle Image Velocimetry Technique

Chen and Fan (1992) developed a Particle Image Velocimetry (PIV) system to provide the 2-D instantaneous flow information in a laser sheet plane. The PIV system has the advantage of being non-intrusive and providing instantaneous as well as averaged flow information. The PIV technique consists of laser sheeting, video recording and image processing. The particle laden flow field is illuminated through the use of a laser sheet and is recorded by a high resolution CCD camera. The selected images of the flow field are digitized through the use of a frame grabber board. The images are then filtered to improve the image quality. A scanning subroutine is executed next, which identifies each particle (liquid tracer, solid particle or a gas bubble) image and computes the centroid of each image. The different phases are then separated into individual sub-directories for later, independent study of each phase. This discrimination of the different phases is based on a prior knowledge of the size distribution of each phase. The processed frames are then scanned for a sequence of preselected velocity vectors using a rapid, logical cross-correlation procedure. The local holdup measurement is performed by counting the pixel numbers occupied in a preselected interrogation region for each sequentially processed frame. This PIV system, therefore, has the capability of simultaneously measuring the full-field, instantaneous flow information, including velocity vectors, holdups and accelerations, for the gas, solid and liquid phases. The instantaneous information obtained by the PIV system can be averaged over many trials to provide the time/volume averaged liquid velocity and gas holdup profiles.

The PIV system was further modified by Reese et al. (1995) to study the 3-D instantaneous flow information. Thus, the macroscopic 3-D flow behavior and associated flow instability could be readily quantified without resorting to the tedious image construction process from 2-D PIV. In order to measure the 3-D flow, correlations of at least two views of the flow field taken from different directions are required. The stereoscopic correlation of a pair of orthogonal images is the most common method to construct the "depth" of the flow field (*e.g.*, Praturi and Brodkey, 1978; Sheu *et al.*, 1982). This technique provides two similar images in which most of the particles visible in one view will also be observable in the other. Although a pair of synchronized cameras located at perpendicular location can be used to record the orthogonal view of the flow field, the alignment of these cameras and the processing of the synchronized data are time-consuming. A system with only a single camera which can simultaneously record the orthogonal images is desirable. Therefore, an optical system (see Fig. 5.1), similar to that designed by Peskin (1972) is developed to create equal optical path lengths for the two views. This optical system provides for the identical parallax in both images and hence simplifies the matching of images.

Images in each orthogonal view are processed and velocity vectors are then mapped. Since the brightness, size and shape of particles can be entirely different for each orthogonal view, the correlation of image pairs can only be based on geometrical consideration. The velocity vectors identified in each orthogonal view are corrected to calculate the 3-D velocity vectors using a set of relations developed by Racca and Dewey (1988). A planar transformation is applied to re-scale the particle images to true dimensions. The instantaneous 3-D velocity vectors, particle size, phase holdups, etc. can therefore be measured for a plane (with finite thickness) in the flow.

Overall Flow Phenomena

Flow visualization in the bulk region of a gas-liquid dispersion and gas-liquid-solid fluidization system was recently studied by Chen *et al.* (1994) using the PIV system. Qualitative observations indicate the existence of three flow conditions over the range of gas flow rates tested in this study. The three flow conditions identified were the dispersed bubble, vortical-spiral flow and turbulent flow. The conventionally defined coalesced bubble regime was divided into vortical-spiral flow at a lower gas velocity and turbulent flow at a high gas velocity. In the dispersed bubble flow regime, the bubble streams are observed to rise rectilinearly with negligible bubble coalescence and relatively uniform size distribution along the column radius. The liquid is carried upward by the bubble induced motion in the vicinity of the ascending bubble streams and falls downward, in a relatively straight manner, between these bubble streams. The dispersed bubble flow regime exists up to a superficial gas velocity of 1.7 cm/sec.

With a further increase in the gas velocity, the bubbles begin to migrate towards the central part of the column and start to form a central bubble stream which moves upward in a rocking, spiral manner. The bubbles, which begin to form the central bubble stream, move upward in clusters with insignificant bubble coalescence up to a superficial gas velocity of 2.1 cm/sec. Bubble coalescence and breakup are observable at superficial gas velocities higher than 2.1 cm/sec, which indicates the beginning of the coalesced bubble regime and the vortical-spiral flow condition. As the gas velocity increases, the rotating frequency of the central bubble stream is intensified which results in a more clearly observable spiral flow pattern of the central bubble stream. The liquid phase is carried upward by the spiral motion of the central bubble stream and flows downward in the same spiral manner between the central bubble stream and the column wall. The liquid flowing downward in this spiral manner also demonstrates a vortical flow pattern with the vortices appearing in the longitudinal direction. A clearer picture of the 3-D vortical-spiral liquid motion and spiral motion of the central bubble stream can be found in Reese *et al.* (1993). The vortical-spiral flow condition, which exists for superficial gas velocities from 2.1 cm/sec to 4.2 cm/sec, is characterized by spiral upward motion of the central bubble stream and by a liquid flow pattern consisting of a spiral upward flow (as carried by the gas phase) in the column center and a downward vortical and spiral pattern in the region between the central bubble stream and the column wall. A proposed general macroscopic flow structure of the vortical-spiral flow condition will be presented later.

The turbulent flow condition is characterized by the existence of large bubbles caused by intensive bubble coalescence which move in a discrete manner and break down the spiral flow pattern of the central bubble stream. This intensive bubble coalescence is observed to occur when the superficial gas velocity exceeds 4.9 cm/sec. The liquid flow pattern in the turbulent flow condition is therefore determined by the bubble wake carriage (Tang and Fan, 1989) and by the drift induced by the rising bubbles (Tsuchiya *et al.*,

1992). Due to the chaotic motion of the liquid phase in the turbulent flow condition, the liquid mixing between the bottom and top of the column is not as rapid as that in the vortical-spiral flow condition.

The regime classification reported in the literature (*e.g.*, Deckwer *et al.*, 1980) usually consists of a coalesced bubble (churn-turbulent) flow regime. Based on the flow mechanisms and flow structures discussed here, however, this coalesced bubble flow regime can be further sub-divided into two flow conditions - the vortical-spiral flow and turbulent flow conditions. The vortical-spiral flow condition has been characterized as a transition condition between the dispersed bubble regime and the turbulent flow. It should be noted that the demarcation criteria for the flow regimes may vary with the design or operating variables such as column size, type of distributor and liquid properties.

Macroscopic Flow Structure of the Vortical-Spiral Flow Condition

The general macroscopic flow structure of a 3-D column operating in the vortical-spiral flow condition is shown in Fig. 5.2. This general macroscopic flow structure is based on the general flow structure of a 2-D fluidization system observed by Tzeng *et al.* (1993), and qualitative observations of the vortical-spiral flow condition. The proposed macroscopic flow structure is confirmed by using the quantitative PIV technique to demonstrate the liquid flow pattern at different radial locations in the column (Chen *et al.*, 1994). The four flow regions of the proposed macroscopic flow structure, as shown in Fig. 5.2, are the descending flow region, the vortical-spiral flow region, the fast bubble flow region and the central plume region.

The descending flow region is located adjacent to the column wall and is characterized by the liquid and/or solid phase streams moving in a straight downward manner. The liquid and/or solid phase streams occasionally demonstrate a spiral downward flow at higher gas velocities in the vortical-spiral flow condition. This region is relatively free of bubbles.

The vortical-spiral flow region is located between the descending flow region and the central bubble stream, and is characterized by downward liquid and/or solid phase streams moving in a spiral and vortical manner. The vortical motion is rather dynamic due to the disturbances caused by the small bubbles entrained in this region and by the larger bubbles rising in the adjacent central bubble stream. The spiral and vortical flow patterns produce good inter-phase mixing characteristics in this region.

The fast bubble flow region is located adjacent to the vortical-spiral flow region and was previously referred to as the central bubble stream. The fast bubble flow region is characterized by significant coalescence and breakup of bubbles which move upward in a spiral manner at a high interstitial velocity. Due to the continuous coalescence and breakup, the bubbles may form more than one spiral bubble streams simultaneously. The number and direction of the spiral bubble streams change dynamically. This region also acts as a baffle to the radial mass transfer of the liquid and/or solid phases between the vortical-spiral flow region and the central plume region.

Located in the column core and surrounded by the fast bubble flow regions is the central plume region. In this region, the bubble size is relatively uniform with less bubble-bubble interactions compared to those in the fast bubble flow region. The central plume region becomes indistinguishable in the 10.2 cm ID column due to the mergence with the fast bubble flow region. This is similar to that observed for a 10.2 cm wide 2-D column (Reese *et al.*, 1993). Further study in scale-up is required for the 3-D column.

The flow structures described above are instantaneous, which may be overlooked when using time-averaging techniques to quantify the flow properties in a 3-D fluidization system. The instantaneous, full-field technique of Particle Image Velocimetry provides the confirmation of these instantaneous flow structures and also can provide the time/volume averaged axial velocity profile along the radial distance to verify the existence of the overall gross circulation.

Chen *et al.* (1994), using the concept that the vortical-spiral flow condition is a transition condition, indicated that the bubble streams in the fast bubble flow region essentially act as a rotating "solid" boundary with axial motion and demonstrated that the flow can be simulated as that between two concentric rotating cylinders of which the inner-cylinder (the fast bubble flow region) is in motion and the outer-cylinder (column wall) is at rest. A thorough study of the states of a single phase flow between two concentric rotating cylinders was conducted by Andereck et al. (1986) who provided flow-regime diagrams for different inner- and outer-cylinder Reynolds numbers at a constant radius ratio (0.883) and an aspect ratio (30). The inner and outer-cylinder Reynolds numbers are defined, respectively, as $R_i=[a(b-a)\Omega_i/\nu]$ and $R_o=[b(b-a)\Omega_o/\nu]$ where ν is the kinematic viscosity of liquid, a and b are the inner- and outer-cylinder radii, and Ω_i and Ω_o are inner- and outer-cylinder angular velocities, respectively. Based on the flow-regime diagrams given by Andereck et al. (1986) and the measured angular velocity of the fast bubble flow region, Chen et al. (1994) showed that the vortical flow structure in the vortical-spiral flow condition is analogous to the turbulent Taylor vortices structure which occurs at R_i greater than 1370 for $R_o=0$. Based on the work of Ludwieg (1964), Chen et al. (1994) also indicated that the spiral motion of the liquid flow in the vortical-spiral flow region is due to the axial motion of the fast bubble flow region.

Entrance Region Flow Phenomena

The instantaneous hydrodynamic behavior in the entrance region of a bubble column has recently been studied by Reese and Fan (1994). Through the use of the PIV system, they found that, in the dispersed bubble regime, the liquid flow in the entrance region is more chaotic and is characterized by higher interstitial velocities than in the bulk region of the column. The chaotic liquid flow in the entrance region is similar to the flow behavior of the bulk region at high gas velocity, although not as intense due to the smaller bubble size. The chaotic behavior of the entrance region, at low gas velocity, may enhance a chemical reaction by providing better mixing or adversely affect the reaction due to over-reaction of the product caused by the increased mixing. In either case, the entrance effects are important and need to be considered in any modeling or design of bubble column reaction systems.

At high gas velocities, the flow develops quickly and the entrance and the bulk regions exhibit similar phenomena. When the gas velocity exceeds 3.0 cm/s the entrance effects become insignificant and only a small area immediately above the distributor is required for the flow and the coherent flow structures to become developed. The flow phenomena of the entrance region was found to be independent of the distributor provided the inlet bubble size was similar and the inlet distribution was uniform.

Using the time/volume averaged results of the liquid velocity profiles at various axial locations, Reese and Fan (1994) concluded that, for all gas velocities studied, one axisymmetric circulation cell is present in the axial direction, which encompasses the entrance and the bulk regions. The velocity distribution and hence the size of the regions remain constant in the bulk region of the column.

PRESSURE AND TEMPERATURE EFFECTS

A knowledge of the effects of pressure and temperature on the hydrodynamics of three-phase fluidization systems is important due to their increased industrial applications in recent years, however, experimental findings of pressure and temperature effects on the hydrodynamics of three-phase systems have not been conclusive. Experimental results of a number of investigators (e.g., Tarmy et al., 1984; Idogawa et al., 1986; De Bruijn et al., 1988; Clark, 1990; Wilkinson and Dierendonck, 1990) indicated that an increased pressure results in an increased gas holdup in bubble columns and three-phase fluidized beds, especially under high gas velocity conditions. Chiba et al. (1989) reported that pressure effects on gas holdup become insignificant when the pressure exceeds a critical value. Experimental evidence (e.g., Idogowa et al., 1987; Wilkinson, 1991) showed that the mean bubble size decreases and the bubble size distribution becomes narrower with an increase in the pressure. In systems with porous plate gas distributors, the bubble size was found to be independent of pressure (Idogowa et al., 1986) and, under low gas velocity conditions, the same independence was found for gas holdup (e.g., Kölbel et al., 1961; Deckwer et al., 1980; Clark, 1990).

The mechanisms of pressure effects on the hydrodynamics of a fluidization system remain inconclusive. Tarmy et al. (1984) attributed the pressure effects to a decreased bubble size due to the increased contribution of gas momentum to the bubble formation process. This implies that a high pressure system operates in the same fashion as a high density system. However, gas density and system pressure were found to affect a system differently as shown by Saberian-Broudjenni et al. (1987) in a three-phase fluidized bed and Clark (1990) in a slurry bubble column. Other pressure dependent properties of gas and liquid phases, e.g., surface tension (Clark, 1990;

Jiang et al., 1992), were proposed to explain pressure effects on gas holdup and bubble size. Wilkinson and Dierendonck (1990) suggested that an increase in gas holdup is due to the decrease in maximum stable wavelength, which results in a smaller maximum stable bubble size in bubble columns.

Little information is available regarding the temperature effects on the general hydrodynamic behavior. Zou et al. (1988) reported that increasing temperature will increase the gas holdup due to a decrease in the surface tension and the liquid viscosity. Furthermore, higher temperature yields higher liquid vapor pressure leading to a significant increase in the net gas flow through the system. However, information on the temperature effects on the hydrodynamic behavior of a bubble column or a three-phase fluidized bed is so limited that no general mechanistic interpretation of the temperature effects has been substantiated.

A major impediment to the advance of the understanding of pressure effects on flow behavior is the lack of experimental results based on the direct flow visualization. Visualization is essential to the quantification of several important hydrodynamic properties such as bubble rise characteristics (i.e., bubble shape, bubble size, etc.), and bubble coalescence and breakup processes. A flow visualization apparatus was developed by Jiang et al. (1994) to study the hydrodynamics of bubble columns and three-phase fluidized beds under high pressure and high temperature conditions. The apparatus can be operated up to a pressure of 21 MPa and a temperature of 180 °C. The flow behavior in the column can be directly visualized through transparent windows. Studies on pressure effects on the bubble behavior and gas holdup in a bubble column (Jiang et al., 1995a), and pressure and temperature effects on initial fluidization in a three-phase fluidized bed (Jiang et al., 1995b) have been conducted in this apparatus.

Pressure Effects on Bubble Size and Gas Holdup in a Bubble Column

Jiang et al. (1995a) investigated the bubble characteristics by means of direct visualization of local flow behavior in a bubble column 5.08 cm in diameter and 80 cm in height (liquid phase: Paratherm heat transfer fluid; gas phase: nitrogen; sparger: multi-orifice ring sparger with orifice diameter of 3 mm). Large bubbles with high rise velocities are frequently observed at low pressures. These large bubbles gradually disappear with an increase in the pressure. Bubble size distributions at pressures of 0.1 MPa and 20.4 MPa are given in Fig. 6.1. These data are measured from the freeze-frame images of bubbles recorded by a video camera located 25 cm above the sparger. It can be seen that the bubble size distribution is much narrower under higher pressure conditions, while the distribution extends over a wide range of bubble size at ambient pressure. Note that although large bubbles exist in the bubble column at low pressures, the number density of these large bubbles is very low. All the number density distributions follow the log-normal distribution. From Fig. 6.1, it can be seen that, it is the tail of the distribution curve that is mostly affected by the pressure. But the bubble size corresponding to the maximum number density in the distribution curve remains almost constant with pressure variation. At a pressure of 20.4 MPa, the number density distribution can also be approximated by the Gaussian distribution due to the disappearance of large bubbles. Similar size distributions and pressure effects were reported by Idogawa et al. (1987) and Wilkinson (1991) at lower pressure ranges for air-water systems. The mean bubble size at high pressures ($P > 1.5$ MPa) is about 1 mm, while the mean bubble size under atmospheric pressure is about 4 mm. The experimental results indicated that the pressure effect on the mean bubble size is insignificant when the pressure is higher than 1.5 MPa.

Figure 6.2 shows the pressure effect on gas holdup in the same bubble column. At low gas velocities, the gas holdup increases almost linearly with the gas velocity for all pressures. Significant pressure effects on the gas holdup are observed at high gas velocity conditions for pressures up to 10 MPa. At pressures higher than 10 MPa, the gas holdup is insensitive to the operating pressure for the gas velocity range examined by Jiang et al. (1995a). Pressure effects on the gas holdup were reported to be significant at high gas velocities by Idogawa et al. (1986) for air-water system with pressures up to 5 MPa. As the pressure increases, bubbles of smaller size are formed with a higher frequency (Kling, 1962; Idogawa et al., 1987a), which results in a higher gas holdup compared with the system operated under atmospheric pressure.

Pressure and Temperature Effects in a Three-Phase Fluidized Bed

The effects of pressure and temperature on the

hydrodynamics of a three-phase fluidized bed, *e.g.*, incipient fluidization, bed expansion and contraction phenomena, etc. were studied by Jiang *et al.* (1995b) via direct flow visualization. The three-phase fluidized bed is of the same geometry as the bubble column in the preceding section. The gas and liquid phases are the same and the solid particles used are glass beads of two different sizes (1 mm and 2.1 mm in diameter).

Incipient Fluidization

Figure 6.3 shows the pressure effect on the liquid minimum fluidization velocity ($U_{mf,l}$) at two gas velocities. It can be seen that $U_{mf,l}$ decreases with an increase in pressure for all gas velocities. The rate of fall of $U_{mf,l}$, however, decreases as the pressure increases. When the pressure increases beyond about 12 MPa, the pressure effects on incipient fluidization almost vanish. To examine the mechanism of the pressure effect on incipient fluidization of a three-phase fluidized bed, Jiang *et al.* (1995b) studied the bubble behavior under incipient fluidization conditions. A series of photographs is given in Fig. 6.4 to represent the bubbles observed under different pressure conditions. As the pressure increases, the mean bubble diameter decreases and, more importantly, large bubbles gradually disappear. Furthermore, when the photographs under different pressures are compared, it can be seen that the difference in bubble size and bubble number density between the first two pressure conditions is much more obvious than that between the last two conditions, although the pressure difference between the former two conditions (3.4 MPa) is much less than that between the latter two conditions (10.6 MPa). The pressure effects on incipient fluidization are attributed to two factors: one is that the liquid viscosity increases with pressure; and the other is that the bubble size reduction and bubble number density increases with pressure.

The temperature effect on the minimum fluidization velocity at a pressure of 2.0 MPa (Jiang *et al.*, 1995b) is shown in Fig. 6.5. For a given gas velocity, the minimum fluidization velocity increases with an increase in temperature. The influence of temperature, however, appears to be less significant for low temperatures, particularly at high gas velocities. Significant temperature effects on the minimum fluidization velocity can be found in a relatively low gas velocity range. Based on the results of Jiang *et al.* (1995b), the temperature effects are mainly due to the decrease of liquid viscosity with temperature.

Bed Expansion

Typical experimental results of pressure effects on bed expansion are shown in Fig. 6.6. It is seen that, upon introduction of a gas into a liquid-solid fluidized bed, substantial bed contraction occurs over a specified pressure range (see Fig. 6.6). The extent of the bed contraction, however, significantly decreases when the pressure is increased [see Figs. 6.6(a) and (b)]. Under given pressure, temperature, and liquid velocity conditions, the bed contraction continues with increasing gas velocity until the bed surface reaches the lowest point beyond which the bed starts to expand with a further increase in gas velocity. The bed expansion starts at a lower gas velocity under high pressure conditions. For example, for 1 mm glass beads, the bed starts to expand at a gas velocity of about 0.8 cm/s at a pressure of 1.46 MPa, while the bed expansion occurs at the gas velocity of 0.5 cm/s at a pressure of 12.84 MPa. Figure 6.6 also shows the effect of pressure on the bed expansion at a temperature of 90°C. As seen in Fig. 6.6(c), no bed contraction occurs at $U_l = 0.24$ cm/s, while a slight bed contraction is observed at $U_l = 0.45$ cm/s when the bed is operated at a pressure of 1.8 MPa. With an increase in pressure, the extent of bed contraction decreases and the bed contraction is no longer observed at pressures higher than 8 MPa for 1 mm glass beads. Compared with the results at a temperature of 23°C, the influence of pressure on the bed expansion is less significant at a temperature of 90°C. The effects of pressure and temperature on the bed expansion can be attributed to the variation of bubble size distribution. When the pressure and temperature increase, both the mean bubble size and the fraction of large bubbles in the bubble size distribution decrease, which diminishes the entrainment of liquid and solids by bubble wakes. Therefore, the extent of bed contraction in a three-phase fluidized bed decreases with an increase in pressure and/or temperature.

CONCLUDING REMARKS

There are similarities in fluid dynamic relationship between bubble rising in liquids and that in liquid-solid suspensions. When the bubble Tadaki numbers are less than 4, bubbles in the liquid-solid fluidized bed are flatter than those in a single fluid. On the other hand, bubbles with Tadaki numbers larger than 10 are less flatter. The deviation of the aspect ratio of the bubble in a fluidized bed from that in a single fluid increases as the particle size and the solids holdup increase. These trends indicate that the dominating force on the bubble between the inertial and the particle collision is alternated at the bubble Tadaki numbers around 4 to 10. The relationship of the bubble rise velocity to the equivalent diameter in the fluidized bed with solids holdup less than 0.4 is similar to that in low viscosity fluids: the bubble rise velocity reaches an asymptotic value when d_e becomes larger than 1.0 cm. The relationship between the bubble Reynolds number and Eötvös number for bubbles in a fluidized bed deviates from that in infinite Newtonian liquids when Eötvös numbers are less than 10. This trend indicates the heterogeneous property of a fluidized bed with respect to the rising bubbles. The correlation proposed in the present study is in an explicit form to predict the bubble rise velocity in a water-glass bead fluidized bed with varying solids holdups and particle sizes.

Bubble wakes play a key role in dictating the transport phenomena of bubbling systems. For example, bubble coalescence takes place through the bubble wake due to the local minimum pressure in the wake. Solids entrainment due to wake carriage of solid particles characterize the freeboard phenomena of a three-phase fluidized bed. The enhancement of the liquid-solid mass transfer and the object-to-emulsion phase heat transfer in a gas-liquid-solid system over that in a liquid-solid system is largely due to the bubble wake effect. Therefore, a clear understanding of the flow structure in the bubble wake is essential.

The macroscopic flow structures in gas-liquid-solid fluidization systems are identified. Specifically, under low gas velocities and high liquid velocities, the dispersed bubble regime prevails as bubbles rise rectilinearly and the liquid/solid phase falls downward between the bubble streams. Further increase in gas velocity results in the transition of flow regime to the vortical-spiral flow condition. In the vortical-spiral flow regime, clusters of bubbles or coalesced bubbles form the central bubble stream moving in a spiral manner with the liquid/solid phase moving in vortical pattern and spiraling downward in a region between central bubble stream and the column wall. The bubble coalescence becomes dominant and forms large bubbles as the flow condition moves into the turbulent flow condition at high gas velocities. Local chaotic motion of the liquid/solid phase, caused by the bubble wake and the drift effects due to the bubble motion, progressively destroys the vortical and spiral flow structure and leads to the turbulent flow structure. The increase in liquid velocity affects the transition of the flow regimes and the size of the flow structure. However, the existence of the solid phase under low solids holdups does not exhibit significant effects on either the transition of the flow regime or the size of the flow structure. The transition of the flow regimes and the flow structure in the vortical-spiral flow region are postulated to be related to the Taylor instability effect characterized by the turbulent vortices structures occurring in the flow between two concentric rotating cylinders.

The operating pressure and temperature affect the fluid dynamic behavior of three-phase fluidized beds, mainly through the variation in bubble characteristics. The minimum fluidization velocity decreases with an increase in pressure and a decrease in temperature. Pressure effects become insignificant when the pressure reaches a certain value and large bubbles disappear. The contribution of reduction in drag forces to the minimum fluidization velocity, however, dominates over the decrease in bubble size with regard to the temperature effect. The extent of bed contraction decreases with an increase in pressure and temperature. Pressure effects on bed expansion are more pronounced in systems at higher temperatures due to reduced liquid viscosity and surface tension, which yield smaller bubbles. The maximum stable bubble size decreases with increasing pressure not only because of the variation in the interfacial physical properties but also due to the gas density variation.

Acknowledgement

The author acknowledges the assistance of Mr. X. Luo, Mr. Shriniwas Chauk, Mr. J. Reese, and Dr. K. Tsuchiya in the preparation of this paper. This work was supported by the National Science Foundation Grant CTS-9200793.

Nomenclature

A_t	total surface area of a circular-cap bubble
a	half width of bubble, inner cylinder radius
a_0	initial value of particle acceleration
b	bubble breadth, major axis of a spherical- or circular-cap bubble, outer cylinder radius
c	constant defined in Eqn. (2.2)
C_p	pressure coefficient
d_e	equivalent bubble diameter: diameter of a sphere having the same volume as the bubble in three

	dimensions (3-D) or diameter of a circle having the same area as the bubble in two dimensions (2-D)
d_p	particle diameter
Eo	Eötvös number (or Bond number, $g\rho_l d_e^2/\sigma$)
Eo_m	Eötvös number based on apparent bed density ($g\rho_m d_e^2/\sigma$)
F	the volume fraction of liquid in a computing cell
g	gravitational acceleration
H	bubble height
H_d	height of a doughnut-shape bubble
h	bubble height; particle penetration depth
h_p	the deepest penetration of a particle
h_i	instantaneous heat transfer coefficient
K_b	parameter accounting for surface conditions (interfacial mobility and/or boundary-layer formation) of a rising spherical bubble, K_b=12,18 and 36 for Hadamard-Rybczynski, Stokes and Levich theories, respectively
Mo	Morton number (or property number) in a gravitational field ($g\mu_l^4/\rho_l\sigma^3$)
n	constant defined in Eqn. (2.2)
P	gas phase pressure
p	liquid phase pressure
$p\text{-}p_\infty$	differential pressure
R	radius of the curvature of a bubble
R_i	Reynolds number based on inner cylinder radius ($R_i=[a(b-a)\Omega_i/\nu]$)
R_o	Reynolds number based on outer cylinder radius ($R_o=[b(b-a)\Omega_o/\nu]$)
Re_b	bubble Reynolds number based on the bubble breadth (bU_b/ν)
Re_e	bubble Reynolds number based on the equivalent bubble diameter ($d_e U_b/\nu$)
Re_t	particle Reynolds number at the terminal velocity ($d_p U_t/\nu$)
T	temperature
Ta	Tadaki number ($Re_e Mo^{0.23}$)
t	time
U_b	bubble terminal velocity, or bubble rise velocity relative to the liquid phase
U_b'	observed rise velocity of a bubble
U_g	superficial gas velocity
U_l	linear liquid velocity
U_ℓ	superficial liquid velocity
$U_{mf,l}$	liquid minimum fluidization velocity
U_{p0}	initial descending velocity of a particle
V_b	volume of bubble
y	lateral displacement from the axis of symmetry of a bubble

Greek Letters

ϵ_l	liquid holdup
ϵ_s	solids holdup
μ_b	apparent bed viscosity
μ_l	liquid viscosity
ν	liquid kinematic viscosity (μ_l/ρ_l)
ρ_m	apparent density of a liquid-solid mixture
ρ_l	liquid density
ρ_s	solid density
σ	surface tension
Ω_i	inner cylinder angular velocity
Ω_o	outer cylinder angular velocity
τ	dimensionless form of time
τ_p	time required to accelerate the particle to U_b

References

Andereck, C. D., S. S. Liu, and H. L. Swinney, "Flow Regimes in a Circular Couette System with Independent Rotating Cylinders," *J. Fluid Mech.*, **164**, 155 (1986).

Baker, C. G. J., E. R. Armstrong, and M. A. Bergougnou, "Heat Transfer in Three-Phase Fluidized Beds," *Powder Technol.*, **21**, 195 (1978).

Bessler, W. F. and H. Littman, "Experimental Studies of Wakes behind Circularly Capped Bubbles," *J. Fluid Mech.*, **185**, 137 (1987).

Bhaga, D. and M. E. Weber, "Bubbles in Viscous Liquids: Shapes, Wakes and Velocities," *J. Fluid Mech.*, **105**, 61 (1981).

Boys, C. V., *Soap Bubble and the Forces which Mould Them*, Doubleday Anchor Book (1959).

Chen, R.C. and L.-S. Fan, "Particle Image Velocimetry for Characterizing the Flow Structure in Three-Dimensional Gas-Liquid-Solid Fluidized Beds," *Chem. Eng. Sci.*, **47**, 3615 (1992).

Chen, R.C., J. Reese, and L.-S. Fan, "Flow Structure in a Three-Dimensional Bubble Column and Three-Phase Fluidized Bed," *AIChE J.*, **40**, 1093 (1994).

Chen, Y.-M. and L.-S. Fan, "Bubble Breakage Mechanisms due to Collision with a Particle in a Liquid Medium," *Chem. Eng. Sci.,* **44**, 117 (1989).

Chiba S, K. Idogowa, Y. Maekawa, H. Moritomi, N. Kato, and T. Chiba, "Neutron Radiographic Observation of High Pressure Three-Phase Fluidization," *Fluidization VI*, J. R. Grace, and L. W. Shemilt, (Ed.), p. 523 (1989).

Chiu, T. M. and E. N. Ziegler, "Heat Transfer in Three-Phase Fluidized Beds," *AIChE J.,* **29**, 677 (1983).

Clark, K. N., "The Effect of High Pressure and Temperature on Phase Distributions in a Bubble Column," *Chem. Eng. Sci.,* **45**, 2301 (1990).

Clift, R., J. R. Grace, and M. E. Weber, *Bubbles, Drops, and Particles*, Academic Press, New York (1978).

Darton, R. C. and D. Harrison, "The Rise of Single Gas Bubbles in Liquid Fluidized Beds," *Trans. I. Chem. Engrs.,* **52**, 301 (1974).

Darton, R. C., "The Physical Behavior of Three-Phase Fluidized Beds," in *Fluidization*, 2nd ed., J. F. Davidson, R. Clift, and D. Harrison, (Ed.), Academic Press, London, p.495 (1985).

Darwin, Sir C., "Note on Hydrodynamics," *Proc. Camb. Phil. Soc.,* **49**, 342 (1953).

Dayan, A. and S. Zalmanovich, "Axial Dispersion and Entrainment of Particles in Wakes of Bubbles," *Chem. Eng. Sci.,* **37**, 1253 (1982).

De Bruijn, T., J., W. J. D. Chase, and W. H. Dawson, "Gas Holdup in a Two-Phase Vertical Tubular Reactor at High Pressure," *Can. J. Chem. Eng.,* **66**, 330 (1988).

Deckwer, W.-D., "On the Mechanism of Heat Transfer in Bubble Column Reactors," *Chem. Eng. Sci.,* **35**, 1341 (1980).

Deckwer, W- D., Y. Louisi, A. Zaida, and M. Ralek, "Hydrodynamic Properties of the Fisher-Tropsch Slurry Process," *Ind. Eng. Chem. Process Des. Dev.,* **19**, 699 (1980).

Deckwer, W.-D., *Reackionstechnik in Blasensäulen*, Otto Salle Varlag GmbH and Co. (1985); (English translation: *Bubble Column Reactors*, Wiley, 1992).

Devanathan, N., D. Moslemian, and M. P. Dudukovic, "Flow Mapping in Bubble Columns Using CARPT," *Chem. Eng. Sci.,* **45**, 2285 (1990).

El-Temtamy, S. A. and N. Epstein,"Simultaneous Solids Entrainment and De-entrainment above a Three-phase Fluidized Bed," in *Fluidization*, J. R. Grace and J. M. Mutsen, (Ed.), Plenum Press, New York, p.519 (1980).

Fan, L.-S., *Gas-Liquid-Solid Fluidization Engineering*, Butterworth, Stoneham, MA (1989).

Fan, L.-S. and K. Tsuchiya, *Bubble Wake Dynamics in Liquids and Liquid-Solid Suspensions*, Butterworth-Heinemann, Stoneham, MA (1990).

Franz, K., T. Borner, H.J. Kantorek, and R. Buchholz, "Flow Structures in Bubble Columns," *Ger. Chem. Eng.,* **7**, 365 (1984).

Grace, J. R., "Shapes and Velocities of Bubbles Rising in Infinite Liquids," *Trans. Inst. Chem. Engrs.,* **51**, 116 (1973).

Harper, R. P., C. W. Hirt, and J. M. Sicilian, "FLOW-3D: Computational Modeling Power for Scientists and Engineers," Flow Science, Inc. report FSI-91-00-1 (1991).

Henriksen, H. K. and K. Ostergaard, "Characteristics of Large Two-Dimensional Air Bubbles in Liquids and in Three-Phase Fluidized Beds," *Chem. Eng. J.,* **7**, 141 (1974).

Hills, J. H., "Radial Non-Uniformity of Velocity and Voidage in a Bubble Column," *Trans. I. Chem. E.,* **52**, 1 (1974).

Hirt, C. W. and B. D. Nichols, "Volume of Fluid (VOF) Method for the Dynamics of Free Boundaries," *J. of Comp. Physics,* **39**, 201 (1981).

Hong, T. and L.-S. Fan, "Bubble Breakup due to Bubble-Particle Collisions," unpublished work,

(1994).

Idogawa, K., K. Ikeda, T. Fukuda, and S. Morooka, "Behavior of Bubbles of Air-Water System in a Column under High Pressure," *Inter. Chem. Eng.*, **26**, 468 (1986).

Idogawa, K., K. Ikeda, T. Fukuda, and S. Morooka, "Effects of Gas and Liquid Properties on the Behaviors of Bubbles in a Column under High Pressure," *Inter. Chem. Eng.*, **27**, 93 (1987).

Inga, J. R., personal communication (1994).

Jiang, P. J., D. Arters, and L.-S. Fan, "Pressure Effects on the Hydrodynamic Behavior of Gas-Liquid-Solid Fluidized Beds," *Ind. Eng. Chem. Res.*, **31**, 2322 (1992).

Jiang, P.J., T.-J. Lin, X. Luo, and L.-S. Fan, "Bed Contraction at High Pressure and Temperature Three-Phase Fluidization," presented at the AIChE Annual Meeting, San Francisco, November 13-18, (1994).

Jiang, P.J., T.-J. Lin, X. Luo, and L.-S. Fan, "Flow Visualization of High Pressure (21 MPa) Bubble Column: Bubble Characteristics," *Trans. I. Chem. Eng.*, **73**, Part A, 269 (1995a).

Jiang, P.J., X. Luo, T.-J. Lin, and L.-S. Fan, "Flow Visualization of High Pressure and High Temperature Three-Phase Fluidization - Incipient Fluidization," presented at Fluidization VIII, Tours, France, May 14-19 (1995b).

Johns, W. F., G. A. Clausen, G. Nongbri, and H. Kaufman, "Texaco T-Star Process for Ebullated Bed Hydrotreating/Hydrocracking," presented at the National Petroleum Refiners Association (NPRA) Conference, San Antonio, March 21-23 (1993).

Kang, Y., I. S. Suh, and S. D. Kim, "Heat Transfer Characteristics of Three Phase Fluidized Beds," *Chem. Eng. Comm.*, **34**, 1 (1985).

Kato, Y., K. Uchida, T. Kago, and S. Morooka, "Liquid Holdup and Heat Transfer Coefficient Between Bed and Wall in Liquid-Solid and Gas-Liquid-Solid Fluidized Beds," *Powder Technol.*, **28**, 173 (1981).

Kato, Y., Y. Taura, T. Kago, and S. Morooka, "Heat Transfer Coefficient Between an Inserted Vertical Tube and a Three-Phase Fluidized Bed," *Kagaku Kogaku Ronbunshu*, **10**, 427 (1984).

Kim, S. D. and A. Laurent, "The State of Knowledge on Heat Transfer in Three-Phase Fluidized Beds," *Int. Chem. Eng.*, **31**, 284 (1991).

Kling, G., "Über die Dynamic der Blasenbildung Beim Begasen von Flüssigkeiten Unter Druck," *Int. J. Heat and Mass Transfer,* **5**, 211 (1962).

Kölbel, H., E. Borchers, and H. Langemann, "Grössenverteilung der Gasblasen in Blasensäulen," *Chem. Eng. Technol.*, **33**, 668 (1961).

Kologerakis, N. and L. A. Behie, "New Generation Three Phase Two Region Bioreactors Used in the Production of Human Vaccines," presented at Fluidization VIII, Tours, France, May 14-19 (1995).

Kumar, S., K. Kusakabe, K. Raghunathan, and L.-S. Fan, "Mechanism of Heat Transfer in Bubbly Liquid and Liquid-Solid Systems: Single Bubble Injection," *AIChE J.*, **38**, 733 (1992).

Kumar, S., K. Kusakabe, and L.-S. Fan, "Heat Transfer in Three-Phase Fluidized Beds Containing Low-Density Particles," *Chem. Eng. Sci.*, **48**, 2407 (1993a).

Kumar, S., K. Kusakabe, and L.-S. Fan, "Three-Phase Fluidized Beds Under High Gas Holdup Conditions," *AIChE J.*, **39**, 1399 (1993b).

Kumar, S. and L.-S. Fan, "Heat-Transfer Characteristics in Viscous Gas-Liquid and Gas-Liquid-Solid Systems," *AIChE J.*, **40**, 754 (1994).

Latif, B.A.J. and J.F. Richardson, "Circulation Patterns and Velocity Distributions for Particles in a Liquid Fluidised Bed," *Chem. Eng. Sci.*, **27**, 1933 (1972).

Larachi, F., G. Kennedy, and J. Chaouki, "A $\dot{\gamma}$-ray Detection System for 3-D Particle Tracking in Multiphase Reactors," *Nucl. Instr. and Meth.*, **A338**, 568 (1994).

Lazarek, G. M. and H. Littman, "The Pressure Field due to a Large Circular Capped Air Bubble Rising in Water," *J. Fluid Mech.*, **66**, 673 (1974).

Lin, J. S., M. M. Chen, and B. T. Chao, "A Novel Radioactive Particle Tracking Facility for Measurement of Solids Motion in Gas Fluidized Beds," *AIChE J.*, **31**, 465 (1985).

Ludwieg, H., "Experimentelle Nachprüfung der Stabilitätstheorien für reibungsfreie Strömungen mit schraubenlinienförmigen Stromlinien," *Z Flugwiss*, **12**, 304 (1964).

Magiliotou, M., Y.-M. Chen, and L.-S. Fan, "Bed-Immersed Object Heat Transfer in a Three-Phase Fluidized Bed," *AIChE J.*, **34**, 1043 (1988).

Miyahara, T., K. Tsuchiya, and L.-S. Fan, "Mechanics of Particle Entrainment in a Gas-Liquid-Solid Fluidized Bed," *AIChE J.*, **35**, 1195 (1989).

Muroyama, K., M. Fukuma, and A. Yasunishi, "Wall-to-Bed Heat Transfer Coefficient in Gas-Liquid-Solid Fluidized Beds," *Can. J. Chem. Eng.*, **62**, 199 (1984).

Muroyama, K., M. Fukuyama, and A. Yasunishi, "Wall-to-Bed Heat Transfer in Liquid-Solid and Gas-Liquid-Solid Fluidized Beds. Part II: Gas-Liquid-Solid Fluidized Beds," *Can. J. Chem. Eng.*, **64**, 409 (1986).

Ostergaard, K., "Fluidization", *Soc. Chem. Ind.*, (London), 58 (1964).

Ostergaard, K., "Flow Phenomena of Three-Phase (Gas-Liquid-Solid) Fluidized Beds," *AIChE Symp. Ser.*, **69**, *128,* 28 (1973).

Page, R. E. and D. Harrison, "Particle Entrainment from a Three-Phase Fluidized Bed," in *Fluidization and Its Applications*, H. Angelino, J. P. Couderc, H. Gibert, and C. Laguerie, (Ed.), Cepadues-Editions, Toulouse, p.393 (1974).

Peskin, C. S., "Flow Patterns around Heart Valves," Ph.D. Thesis, Albert Einstein College of Medicine, Yeshiva University, New York (1972).

Praturi, A.K. and R.S. Brodkey, "A Stereoscopic Visual Study of Coherent Structures in Turbulent Shear Flow," *J. Fluid Mech.*, **89**, 251 (1978).

Racca, R. G. and J. M. Dewey, "A Method for Automatic Particle Tracking in a Three-Dimensional Flow Field," *Exp. Fluids*, **6**, 25 (1988).

Raghunathan, K., S. Kumar, and L.-S. Fan, "Pressure Distribution and Vortical Structure in the Wake Behind Gas Bubble in Liquid and Liquid-Solid Systems," *Int. J. Multiphase Flow,* **18**, 41 (1992).

Ramachandran, P. A. and R. V. Chaudhari, *Three-Phase Catalytic Reactors*, Gordon and Breach Science Publishers, (1983).

Reese, J., R.C. Chen, J.-W. Tzeng, and L.-S. Fan, "Characterization of the Macroscopic Flow Structure in Gas-Liquid and Gas-Liquid-Soli Fluidization Systems Using Particle Image Velocimetry," *Int. Video J. Eng. Research*, **3**, 17 (1993).

Reese, J. and L.-S. Fan, "Transient Flow Structure in the Entrance Region of a Bubble Column using Particle Image Velocimetry," *Chem. Eng. Sci.*, **49**, 5623 (1994).

Reese, J., R.C. Chen, and L.-S. Fan, "Three-Dimensional Particle Image Velocimetry for use in Three-Phase Fluidization Systems," *Exp. Fluids*, in press (1995).

Rietema, K., "Science and Technology of Dispersed Two-Phase Systems - I and II," *Chem. Eng. Sci.*, **37**, 1125 (1982).

Saberian-Broudjenni, M., G. Wild, J.-C. Charpentier, Y. Fortin, J.-P. Euzen, and R. Patoux,

"Contribution to the Hydrodynamics Study of Gas-Liquid-Solid Fluidized Bed Reactors," *Int. Chem. Eng.*, **27**, 423-440 (1987).

Shah, Y. T., *Gas-Liquid-Solid Reactor Design*, McGraw-Hill (1979).

Sheu, Y.-H., T.P.K. Chang, G.B. Tatterson, and D.S. Dickey, "A Three-Dimensional Measurement Technique for Turbulent Flow," *Chem. Eng. Comm.*, **17**, 67 (1982).

Song, G.-H., "Hydrodynamics and Interfacial Gas-Liquid Mass Transfer of Gas-Liquid-Solid Fluidized Beds," Ph.D. Thesis, The Ohio State Univ., Columbus, OH (1989).

Stewart, P. S. B. and J. F. Davidson, "Three-Phase Fluidization: Water, Particles and Air," *Chem. Eng. Sci.*, **19**, 319 (1964).

Suh, I.-S., G. T. Jin, and S. D. Kim, "Heat Transfer Coefficients in Three Phase Fluidized Beds," *Int. J. Multiphase Flow.*, **11**, 255 (1985).

Suh, I.-S. and W.-D. Deckwer, "Unified Correlation of Heat Transfer Coefficients in Three-Phase Fluidized Beds," *Chem. Eng. Sci.*, **44**, 1455 (1989).

Tang, W.-T. and L.-S. Fan, "Hydrodynamics of a Three-phase Fluidized Bed Containing Low Density Particles," *AIChE J.* **35**, 355 (1989)

Tarmy, B., M. Chang, C. Coulaloglou, and P.Ponzi, "Hydrodynamic Characteristics of Three-Phase Reactors," *Chem. Eng.*, Oct., 18 (1984).

Tarmy, B.L. and C. A. Coulaloglou, "Alpha - Omega and Beyond: Industrial View of Gas-Liquid-Solid Reactor Development," *Chem. Eng. Sci.*, **47**, 3231 (1992).

Tsuchiya, K. and L.-S. Fan, "Near-Wake Structure of a Single Gas Bubble in a Two-Dimensional Liquid-Solid Fluidized Bed: Vortex Shedding and Wake Size Variation," *Chem. Eng. Sci.*, **43**, 1167 (1988).

Tsuchiya, K., G.-H. Song, and L.-S. Fan, "Effects of Particle Properties on Bubble Rise and Wake in a Two-Dimensional Liquid-Solid Fluidized Bed," *Chem. Eng. Sci.*, **45**, 1429 (1990).

Tsuchiya, K., G.-H. Song, W.-T. Tang, and L.-S. Fan, "Particle Drift Induced by a Rising bubble in a Liquid-Solid Fluidized Bed Containing Low-Density Particles," *AIChE J.*, **38**, 1847 (1992).

Tsutsumi, A., J.-Y. Nieh, and L.-S. Fan, "Particle Wettability Effects on Bubble Wake Dynamics in Gas-Liquid-Solid Fluidization," *Chem. Eng. Sci.*, **46**, 2381 (1991).

Tzeng, J.-W., R. C. Chen, and L.-S. Fan, "Visualization of Flow Characteristics in a 2-D Bubble Column and Three-Phase Fluidized Bed," *AIChE J.*, **39**, 733 (1993).

Ulbrecht, J.J., Y. Kawase, and K.F. Auyeung, "More on Mixing of Viscous Liquids in Bubble Columns," *Chem. Eng. Comm.*, **35**, 175 (1985).

Vakhrushev, I. A. and G. I. Efremov, *Chem. Technol. Fuels Oils (USSR)*, **5/6**, 376 (1970).

Viswanathan, S., A. S. Kakar, and P. S. Murti, "Effect of Dispersed Bubbles into Liquid Fluidized Beds with Gas-Liquid Cocurrent Upflow," *Chem. Eng. Sci.*, **19**, 903 (1964).

Walter, J.F. and H.W. Blanch, "Liquid Circulation Patterns and Their Effect on Gas Hold-up and Axial Mixing in Bubble Columns," *Chem. Eng. Comm.*, **19**, 243 (1983).

Wasan, D T. and M. S. Ahluwalia, "Consecutive Film and Surface Renewal Mechanism for Heat or Mass Transfer from a Wall," *Chem. Eng. Sci.*, **24**, 1535 (1969).

Wilkinson, P. M. and L. L. V. Dierendonck, "Pressure and Density Effects on Bubble Breakup and Gas Holdup in Bubble Columns," *Chem. Eng. Sci.*, **45**, 2309 (1990).

Wilkinson, P. M., "Physical Aspects and Scale-up of High Pressure Bubble Columns," Ph.D. Thesis, University of Groningen, The Netherlands (1991).

Yang, Y. B., N. Devanathan, and M. P. Dudukovic, "Liquid Backmixing in Bubble Columns via Computer-Automated Radioactive Particle Tracking (CARPT)," *Exp. Fluids*, **16**, 1 (1993).

Zhang, J., E. Lee, and L.-S. Fan, "The Bubble Rise Characteristics and Bed Rheology in Liquid-Solid Fluidized Beds," unpublished work, (1994).

Zou, R., X. Jiang, B. Li, Y. Zu, and L. Zhang, "Studies on Gas Holdup in a Bubbble Column Operated at Elevated Temperatures," *Ind. Eng. Chem. Res.*, **27**, 1910 (1988).

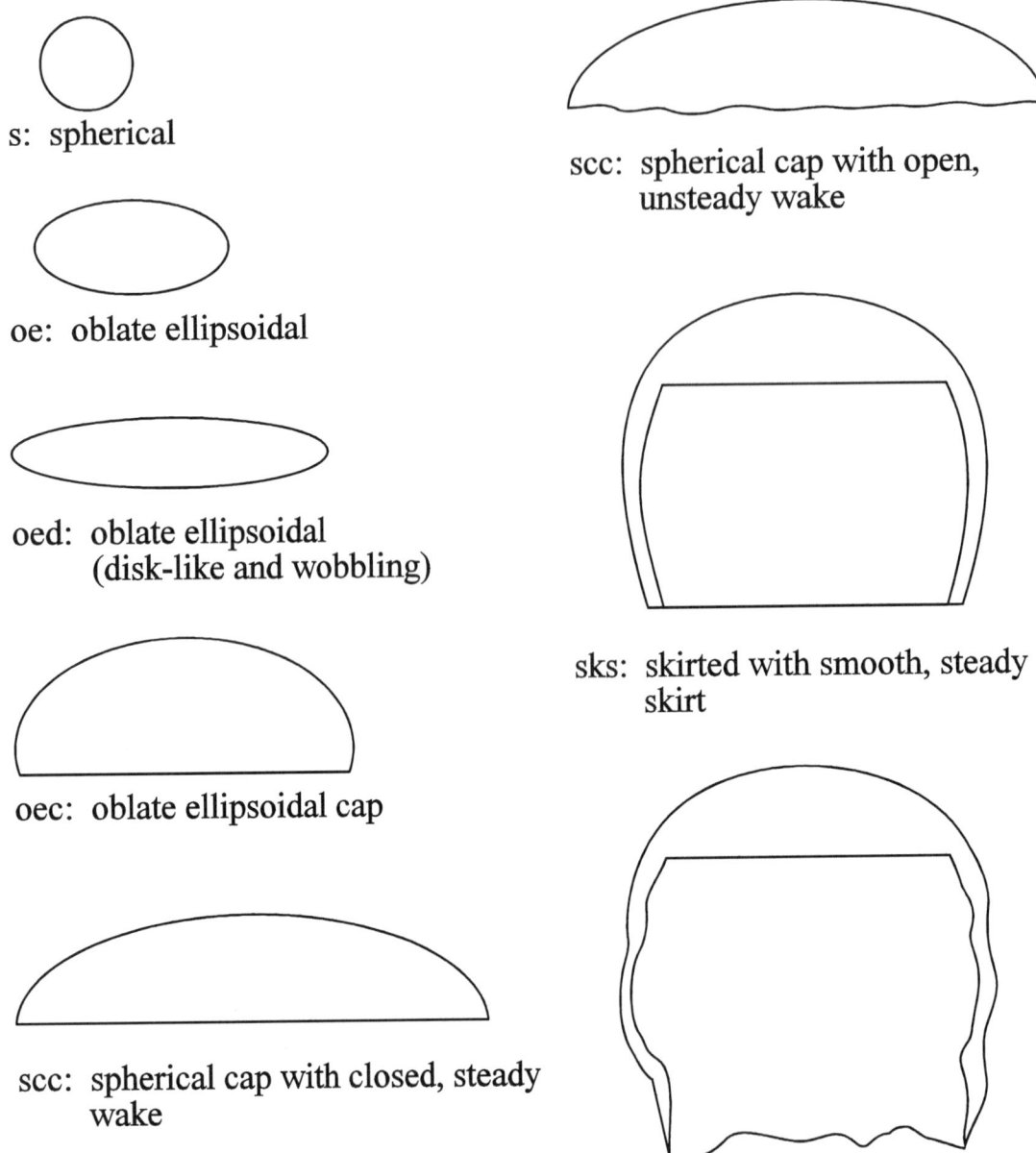

Figure 2.1 Sketches of various bubble shapes observed in infinite Newtonian liquids (after Bhaga and Weber, 1981).

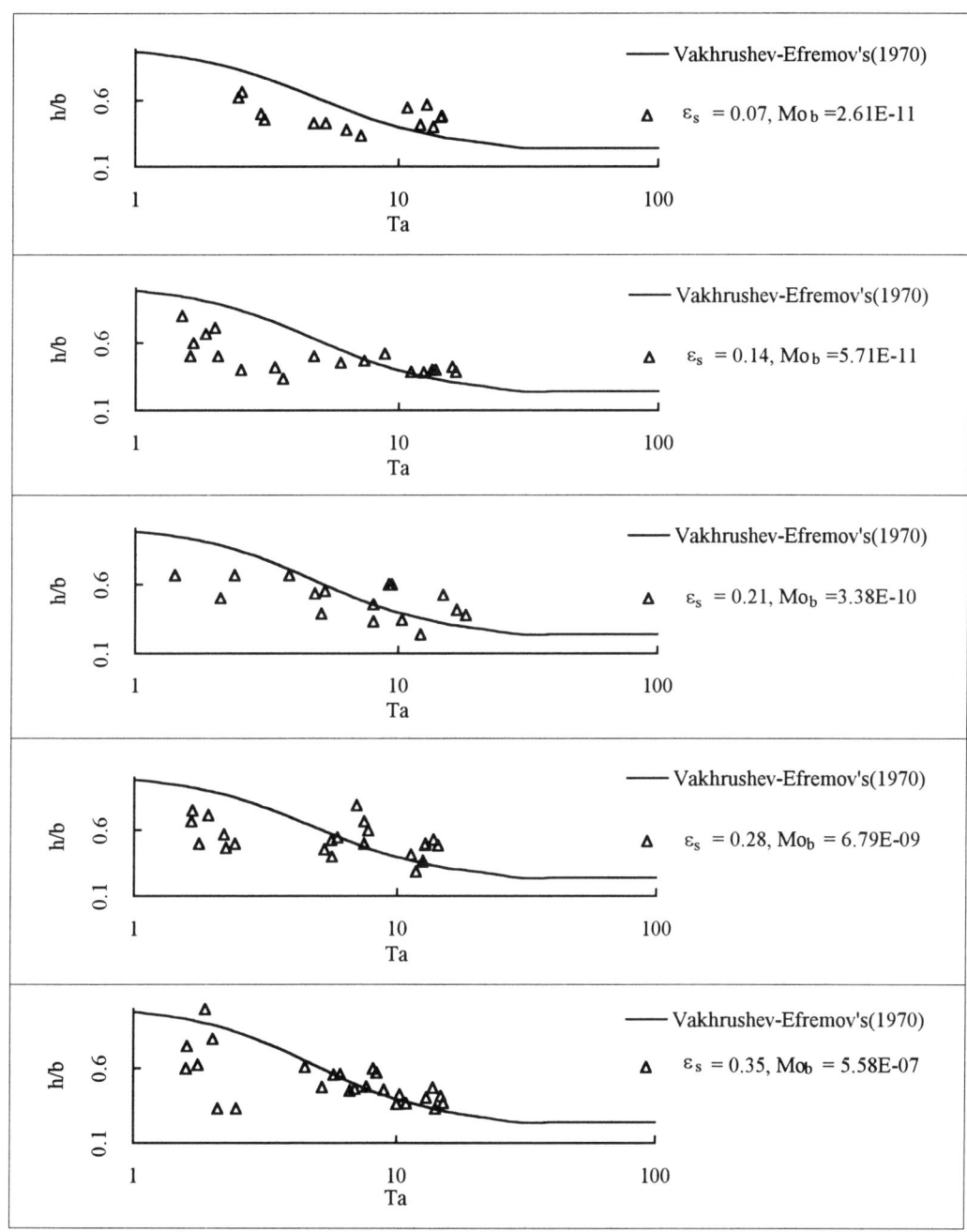

Figure 2.2 Bubble aspect ratio in a three-dimensional water-glass bead fluidized bed with different solid holdups.

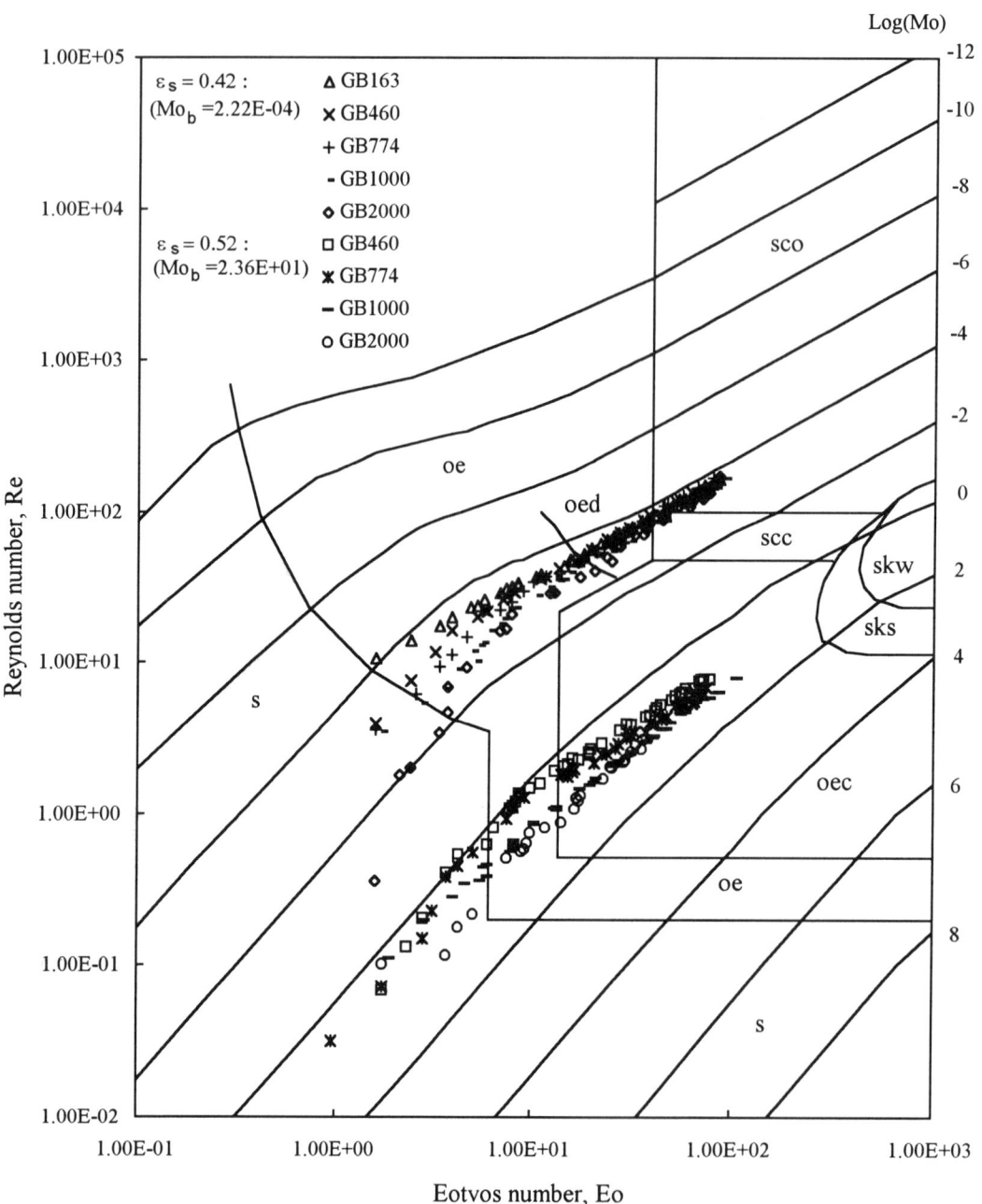

Figure 2.3 Comparison of single bubble rise characteristics in a water-glass bead fluidized bed (data from Jang, 1989) with that in infinite Newtonian liquids.

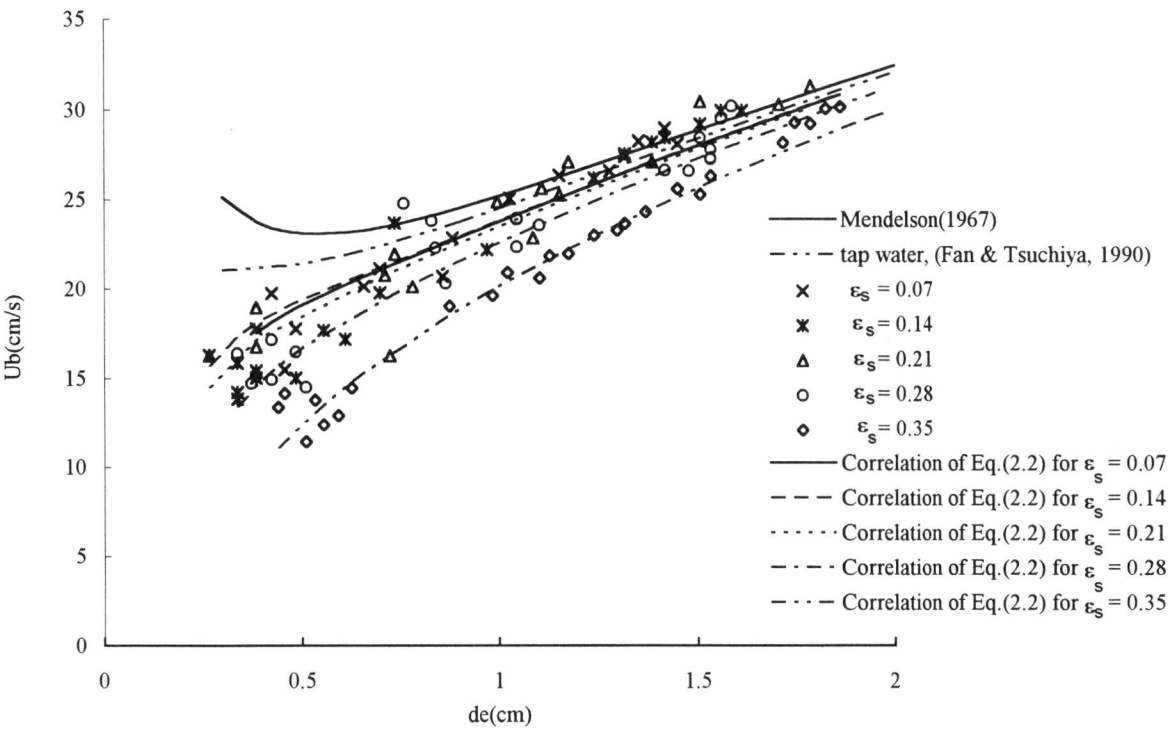

Figure 2.4 Bubble rise velocity in a water-1 mm glass bead fluidized bed at low to intermediate solids holdups.

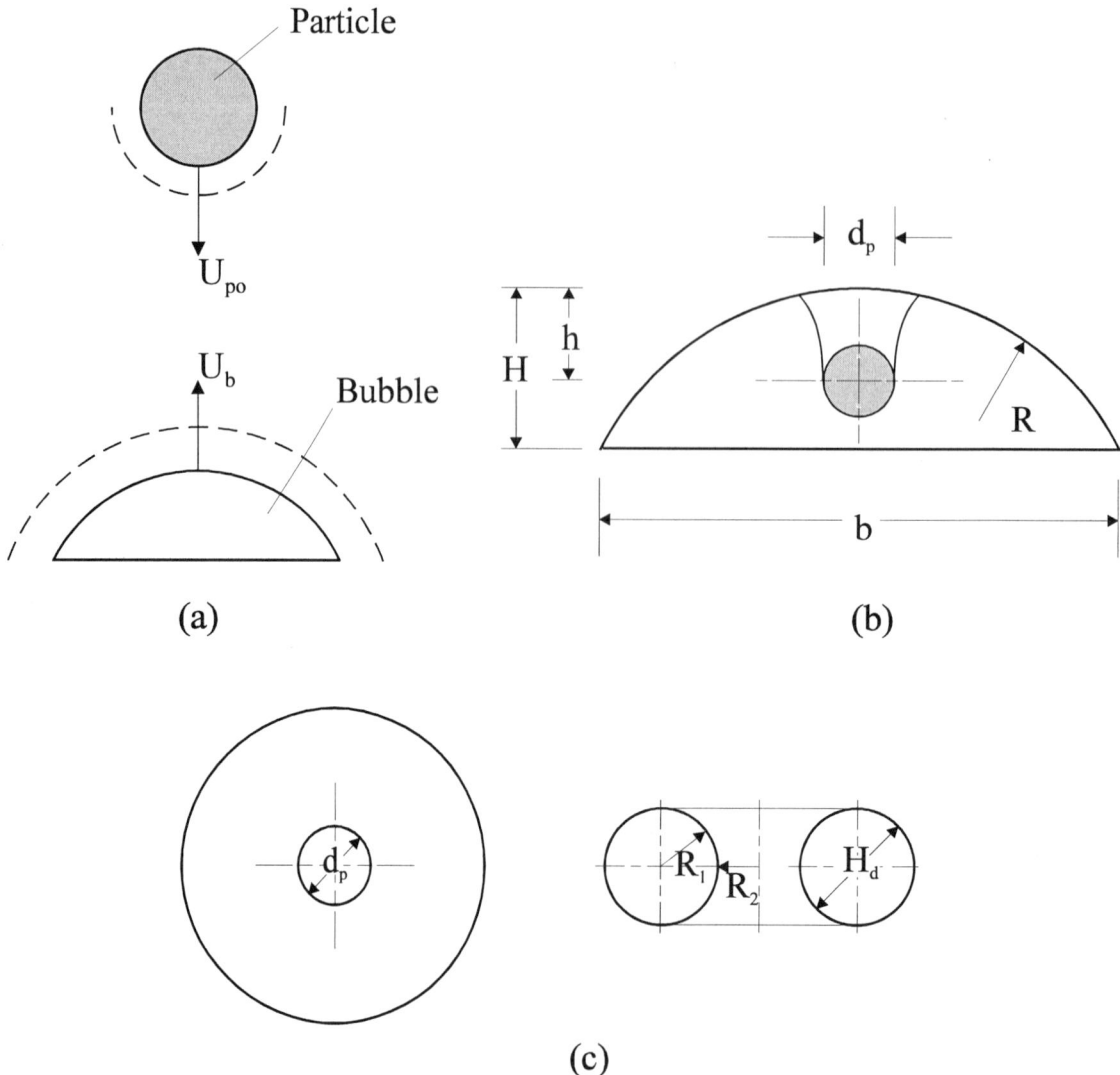

Figure 3.1 (a) Configuration of a particle approaching a spherical-cap bubble.
(b) Configuration of a particle colliding with a spherical-cap bubble.
(c) Configuration of a doughnut-shape bubble (Chen and Fan, 1989).

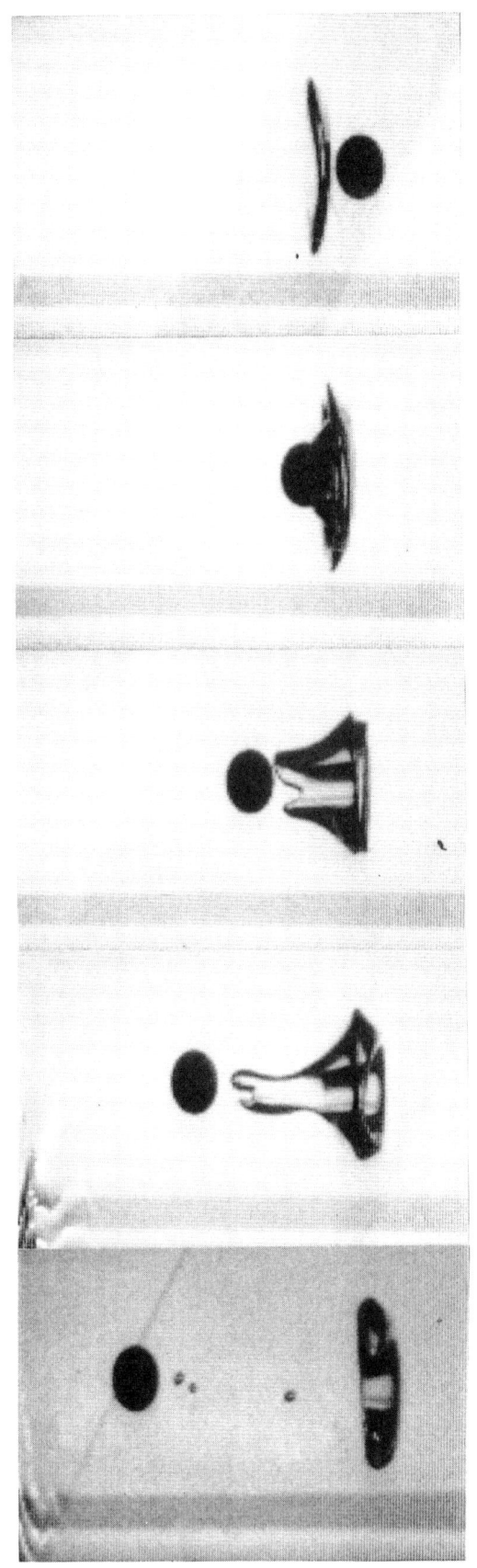

Figure 3.2 (a) Collision sequence of a copper sphere ($d_p = 1.22$ cm) and a spherical-cap bubble ($d_e = 1.97$ cm) in 80% glycerine solution.

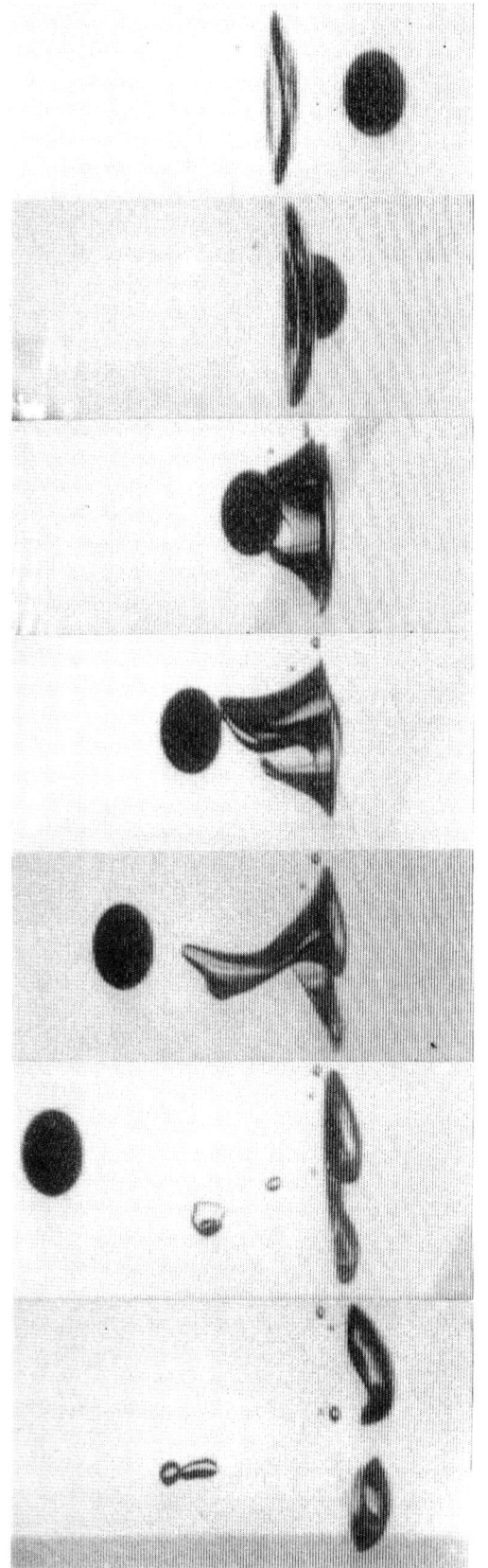

Figure 3.2 (b) Collision sequence of a copper sphere ($d_p = 1.53$ cm) and a spherical-cap bubble ($d_e = 1.97$ cm) in 80% glycerine solution.

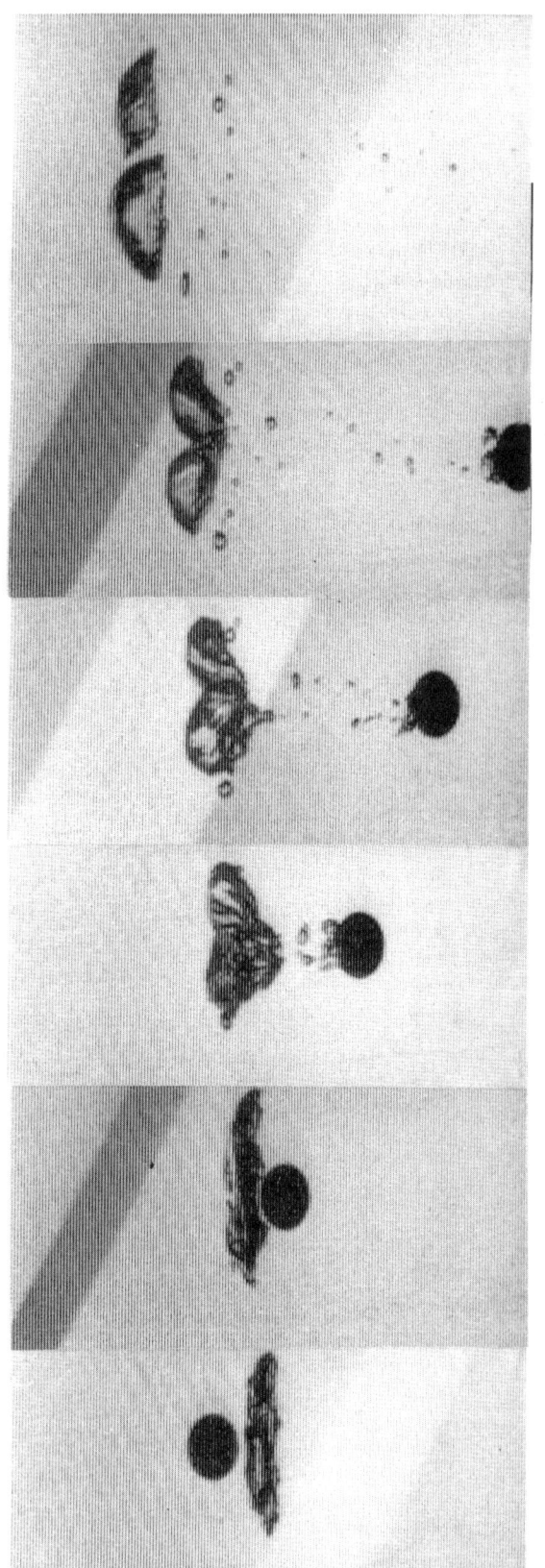

Figure 3.4 Collision sequence of a copper sphere ($d_p = 1.22$ cm) and a spherical-cap bubble ($d_e = 1.97$ cm) in water.

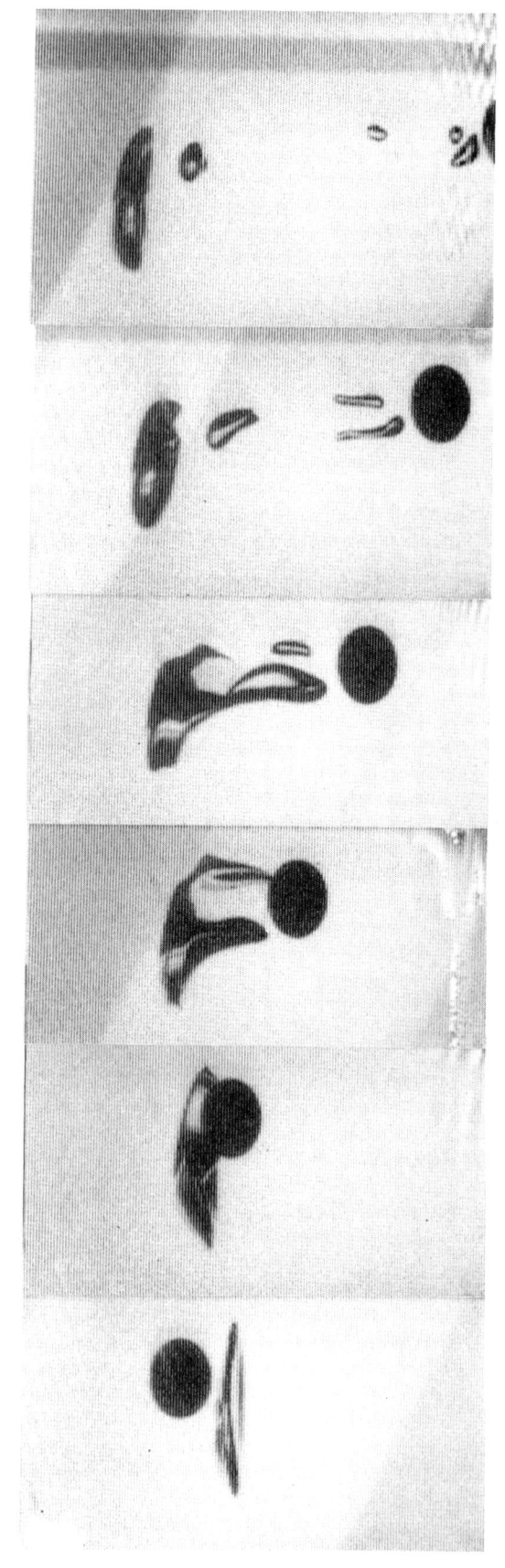

Figure 3.3 Collision sequence of a copper sphere ($d_p = 1.53$ cm) and a spherical-cap bubble ($d_e = 1.97$ cm) in 80% glycerine solution.

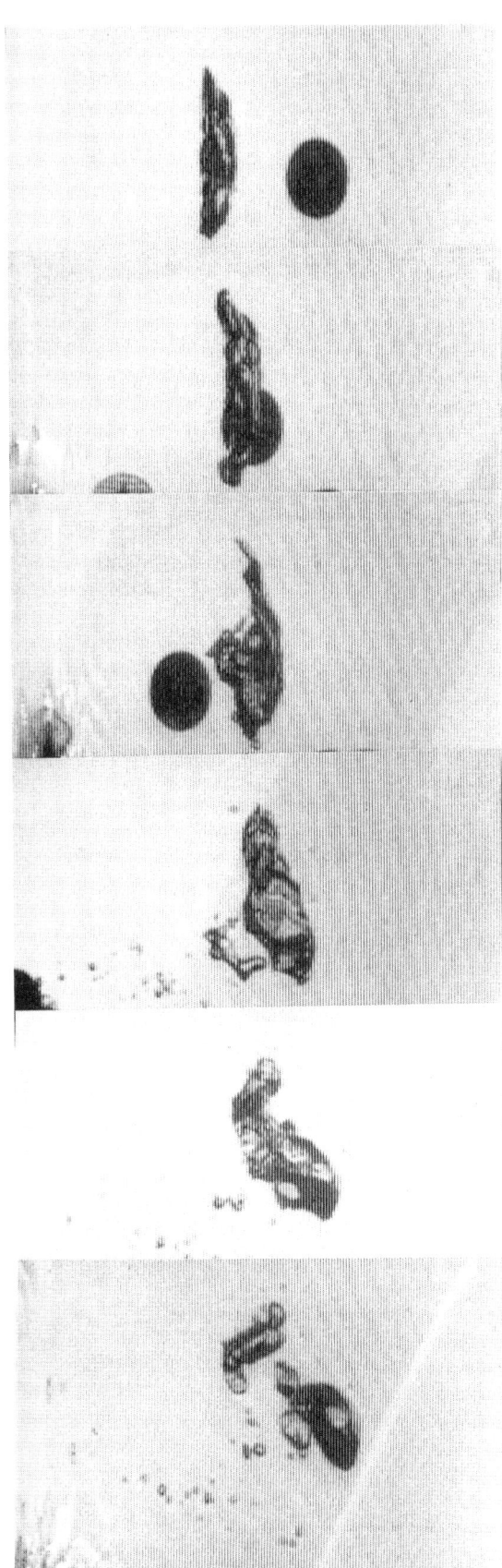

Figure 3.5　Collision sequence of a copper sphere (d_p=1.53 cm) and a spherical-cap bubble (d_e=1.97 cm) in water.

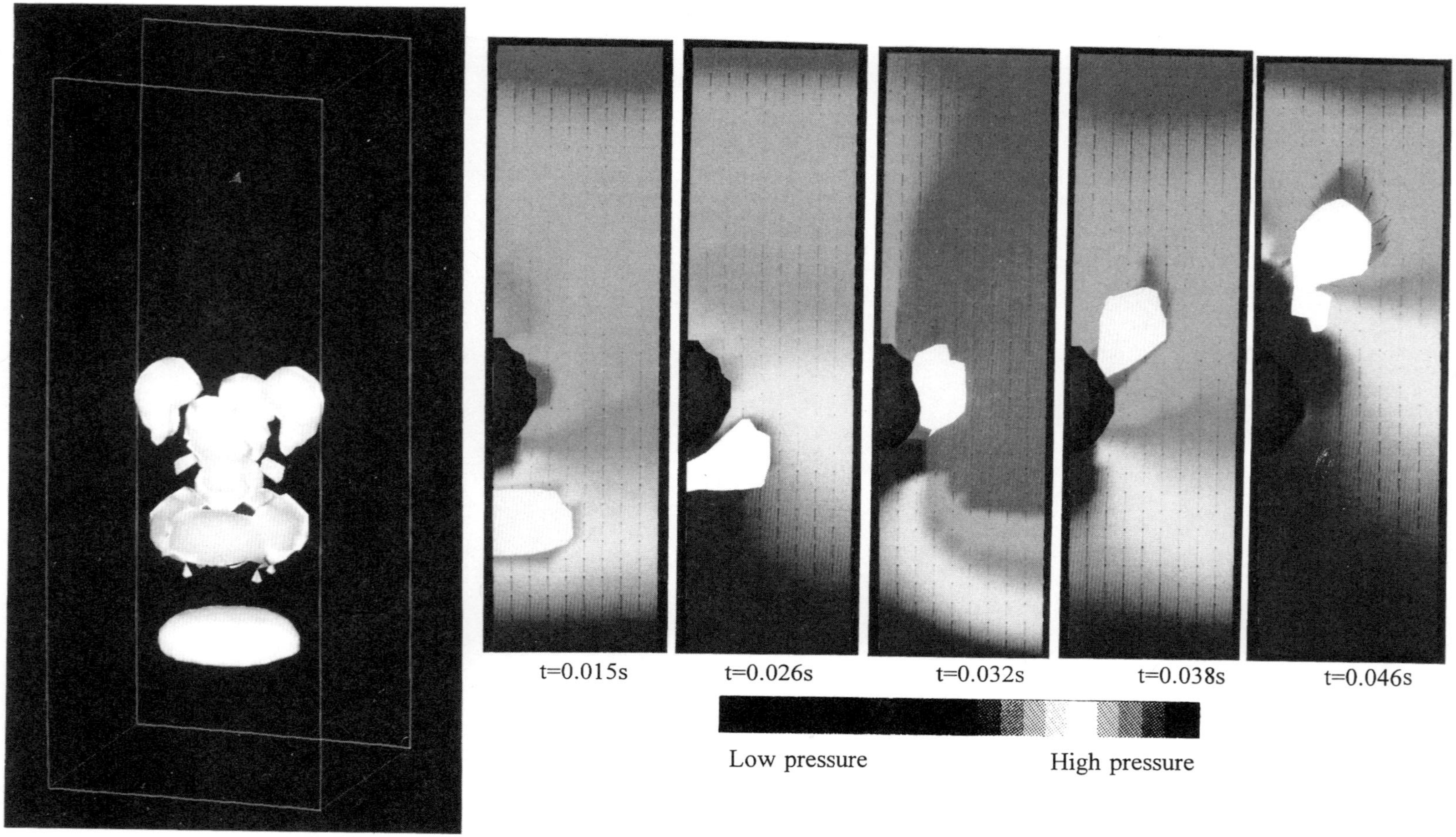

Figure 3.6 Collision sequence of a copper sphere ($d_p = 1.1$ cm) and a spherical-cap bubble ($d_e = 2.0$ cm) in water from the numerical simulation.

t=0.015s
t=0.026s
t=0.032s
t=0.038s
t=0.046s

Low pressure
High pressure

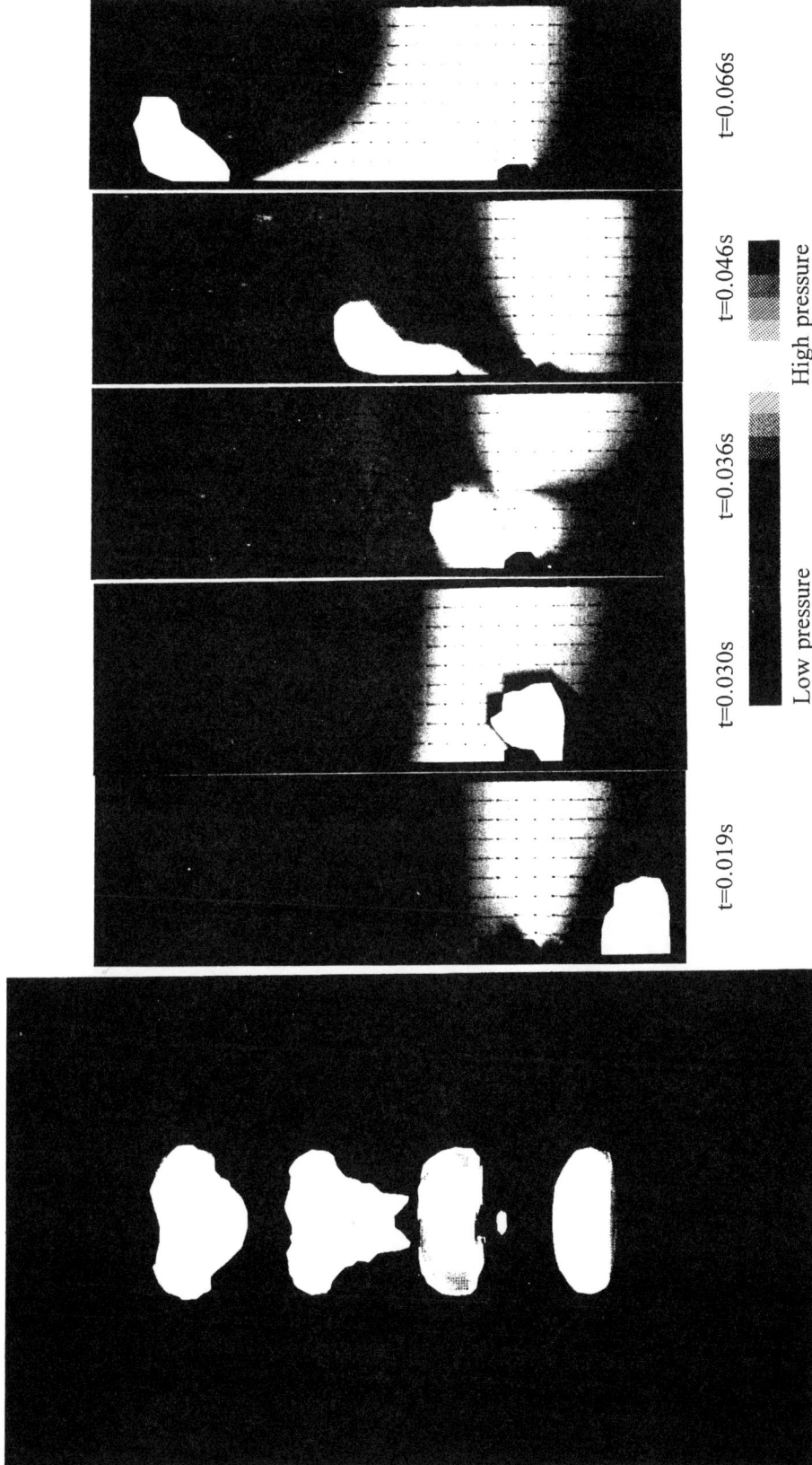

Figure 3.7 Collision sequence of a copper sphere ($d_p = 0.6$ cm) and a spherical-cap bubble ($d_e = 2.0$ cm) in water from the numerical simulation.

Figure 4.1 Relationship between the vortical structure and the pressure signal. Visualization performed in the freeboard region of 0.163 mm beads with the entrained particles serving as tracers (Reprinted from *Int. J. Multiphase Flow,* vol. 18, Raghunathan *et al.*, "Pressure Distribution and Vortical Structure in the Wake Behind Gas Bubble in Liquid and Liquid-Solid Systems", p. 41, 1992©, with kind permission from Elsevier Science Ltd., UK).

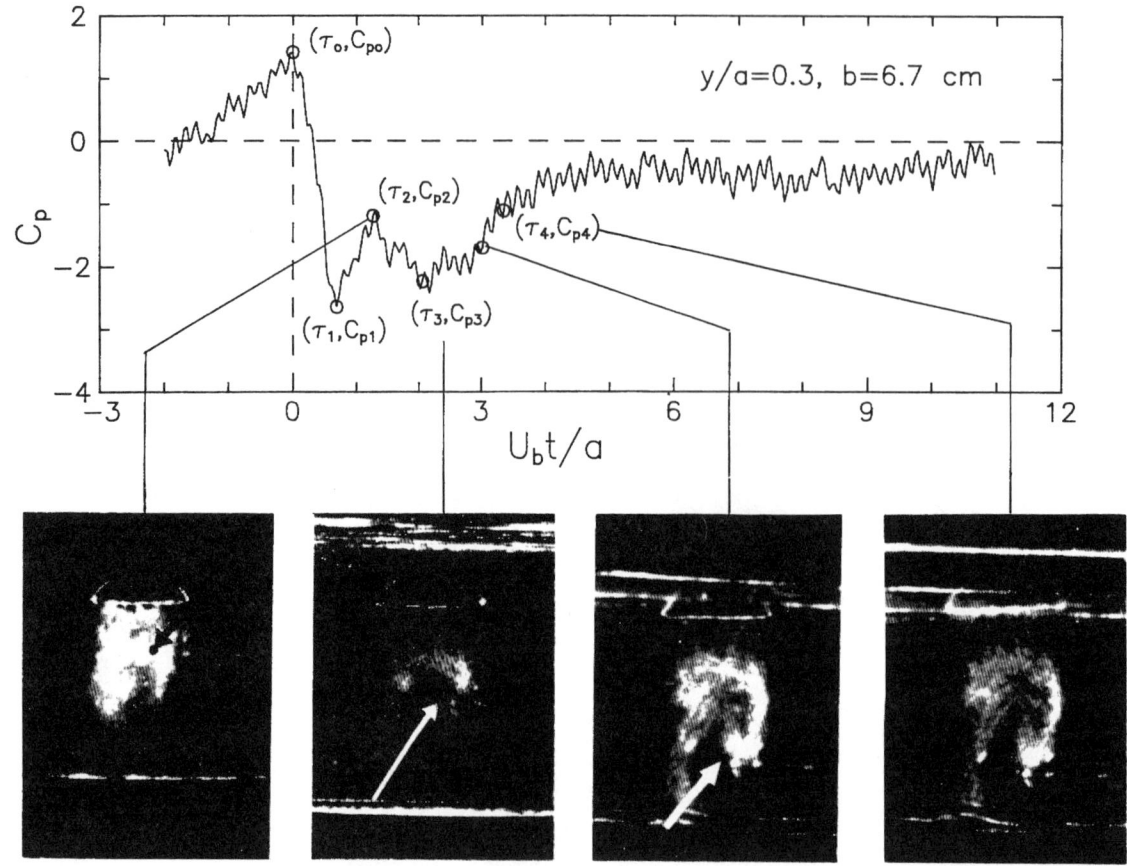

Figure 4.2 Circular-cap bubble, with its primary wake being visualized through entrained particles, and associated particle drift in a water-calcium alginate particle fluidized bed (after Tsuchiya *et al.*, 1992).

Figure 4.3 Instantaneous heat transfer data due to bubble passage synchronized with visualization for probe located at the center of column in a liquid-solid fluidized bed of low-density gel beads (Reprinted from *Chem. Eng. Sci.*, vol. 48, Kumar *et al.*, "Heat Transfer in Three-Phase Fluidized Beds Containing Low-Density Particles", p. 2407, 1993©, with kind permission from Elsevier Science Ltd., UK).

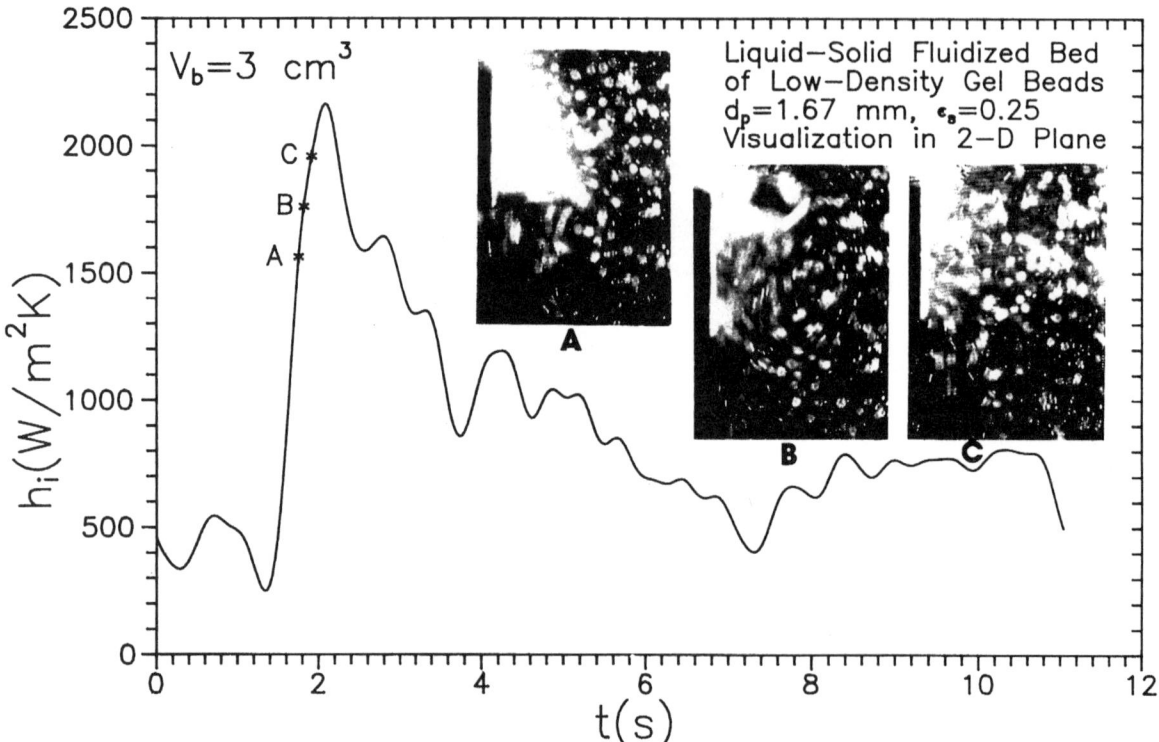

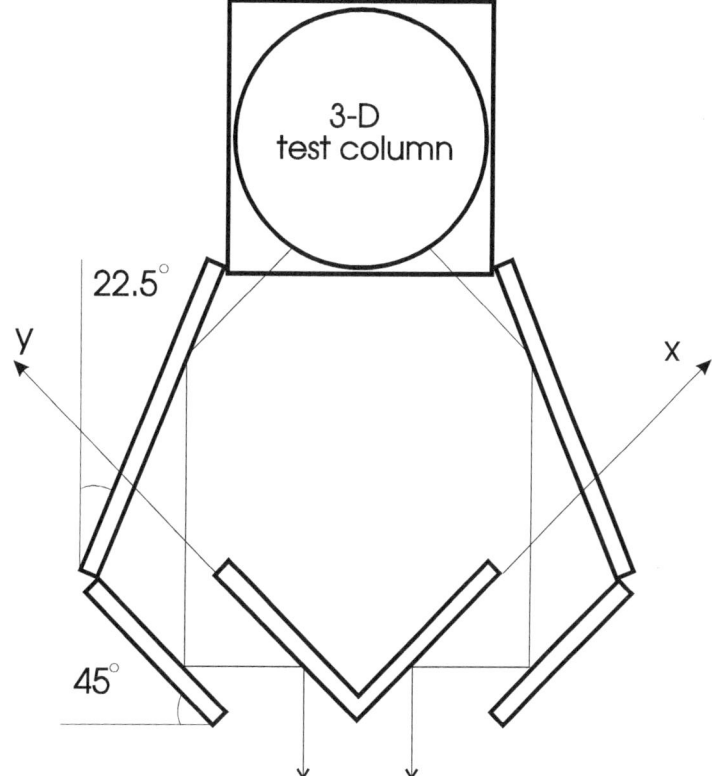

Figure 5.1 Top view of mirror arrangement to produce side by side orthogonal views for one camera 3-DPIV.

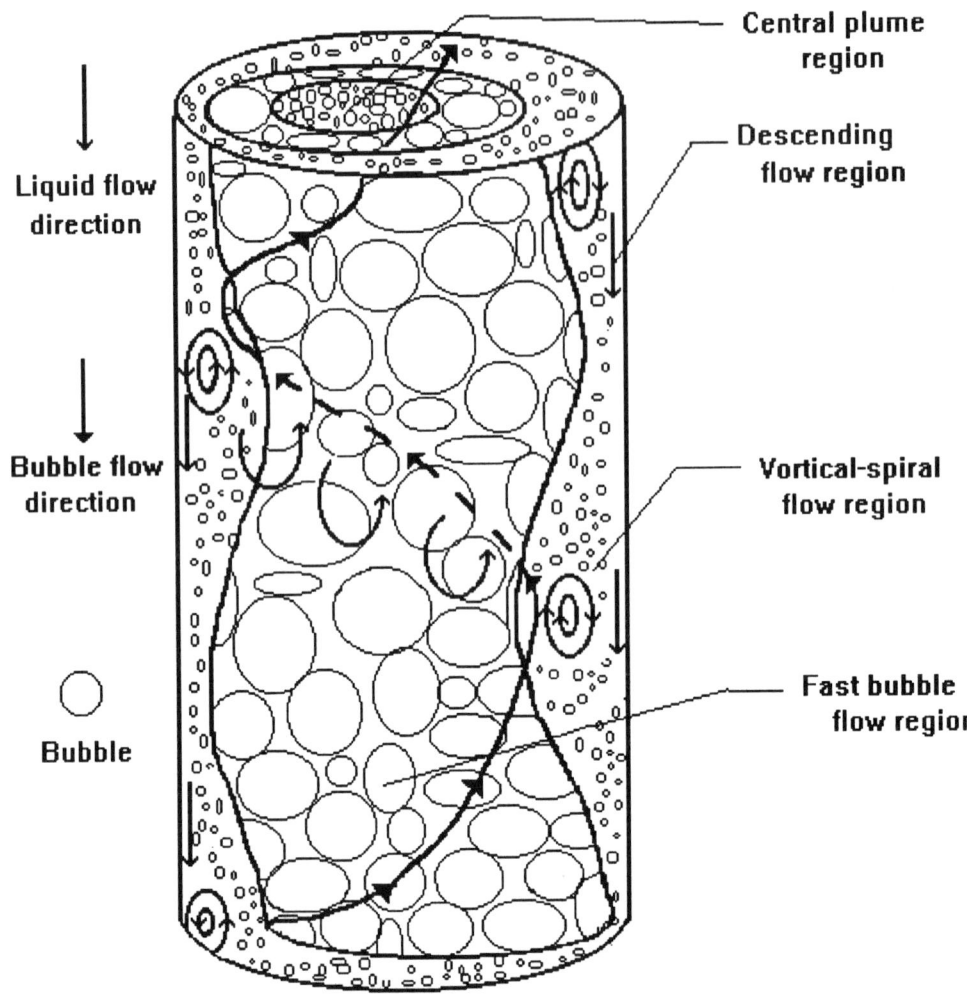

Figure 5.2 Macroscopic flow structure in the vortical-spiral flow condition in a three-dimensional gas-liquid or gas-liquid-solid system (Chen et al., 1994).

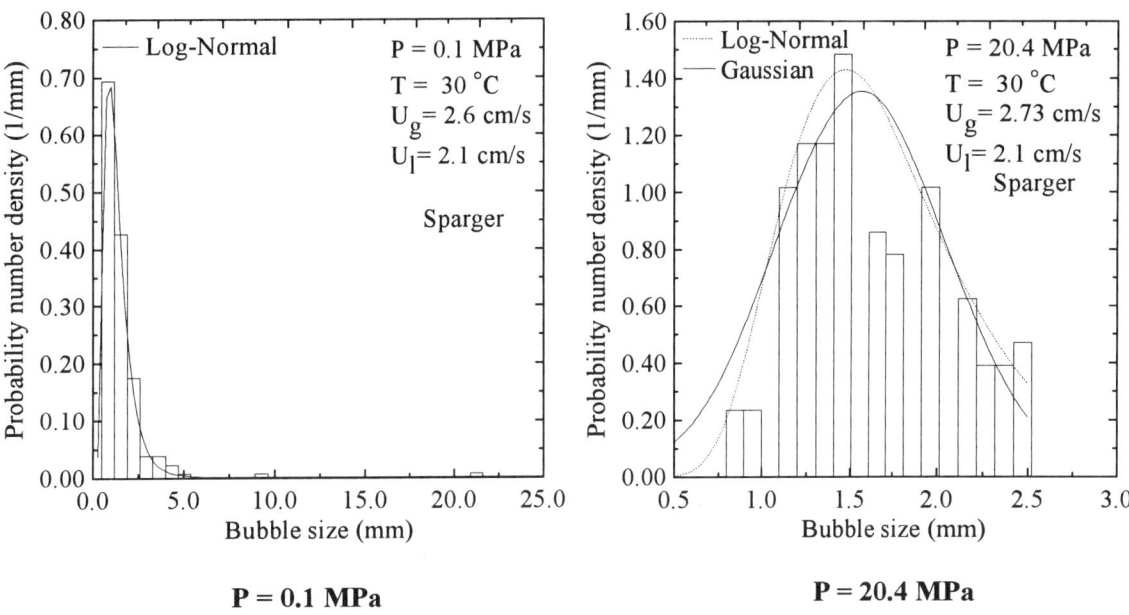

Figure 6.1 Pressure effects on the local bubble size distribution in a bubble column (Jiang et al., 1995a).

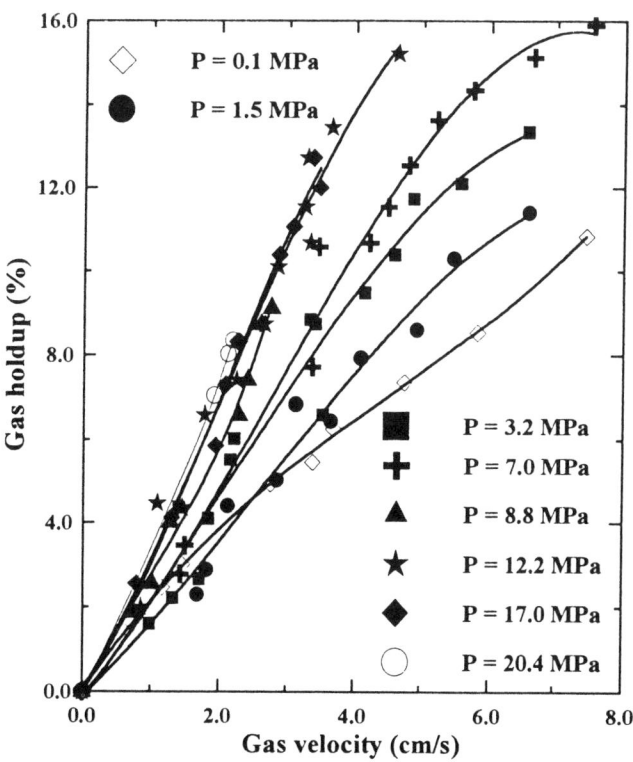

Figure 6.2 Pressure effects on the gas holdup in a bubble column (Jiang et al., 1995a).

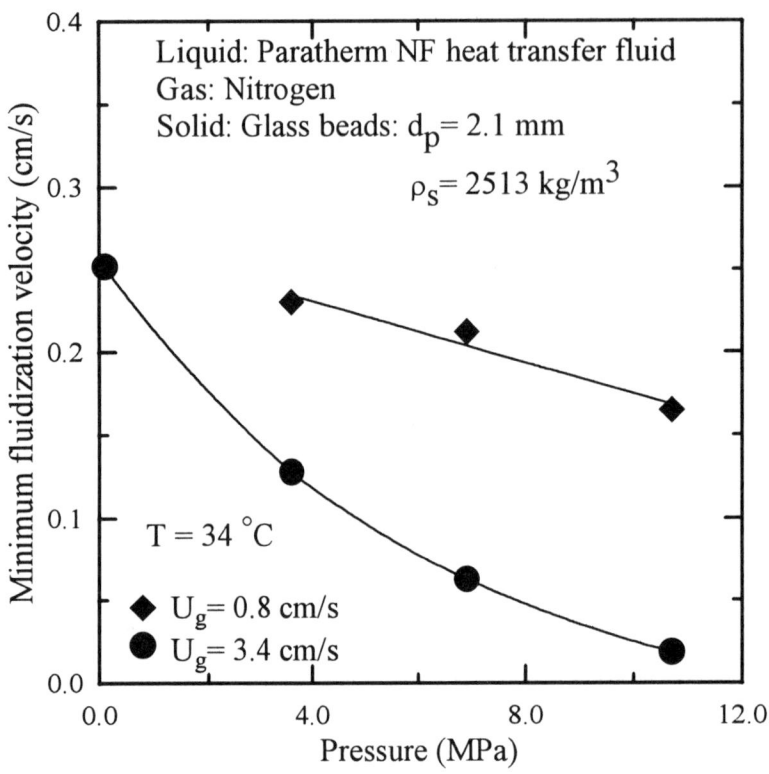

Figure 6.3 Pressure effects on the minimum fluidization velocity at a temperature of 34°C (Jiang et al., 1995b).

Figure 6.4 Photographs of bubbles emerging from bed surface under incipient fluidization conditions for 2.1 mm glass beads at various pressures (T=54°C) (Jiang *et al.*, 1995b).

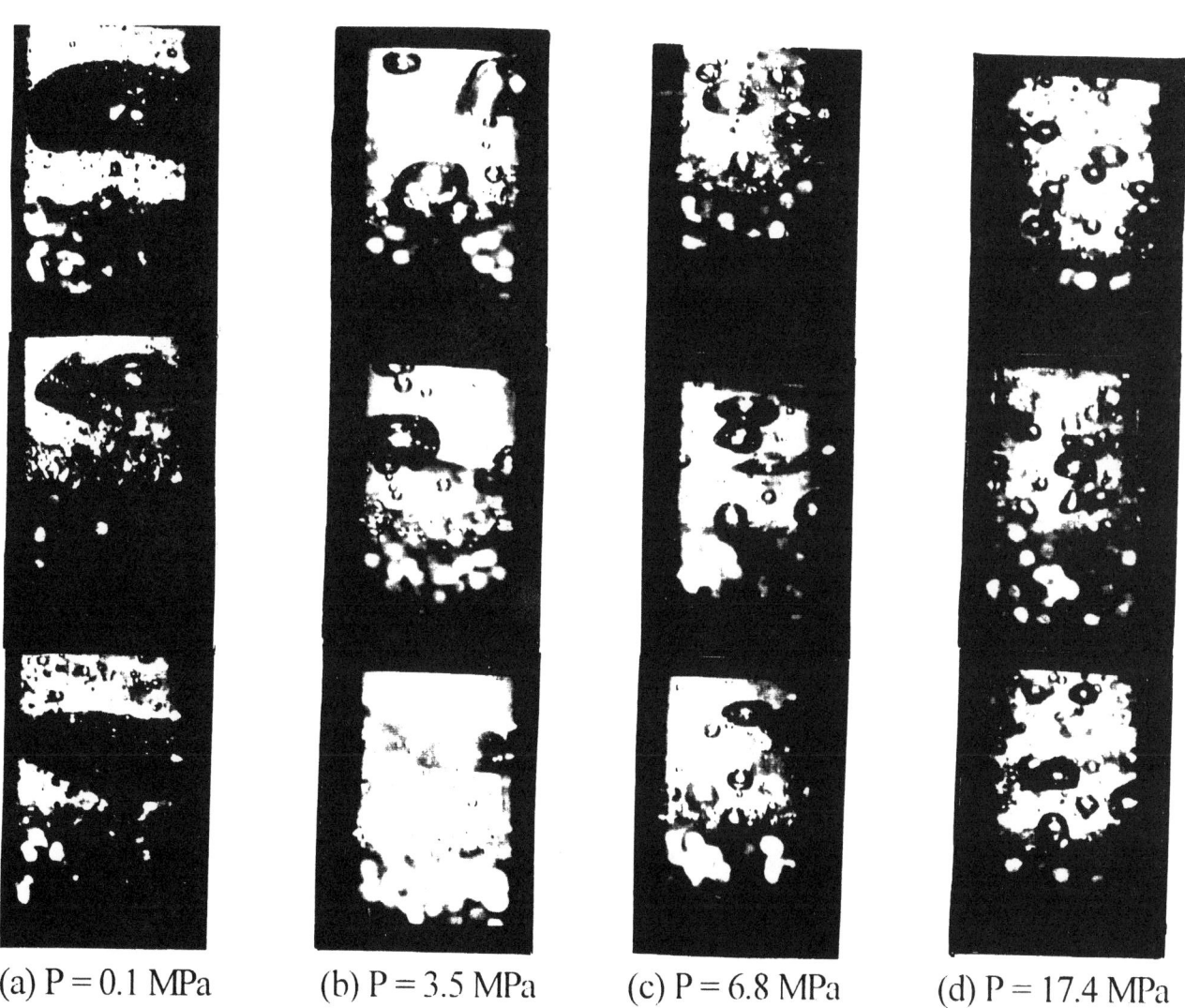

(a) P = 0.1 MPa (b) P = 3.5 MPa (c) P = 6.8 MPa (d) P = 17.4 MPa

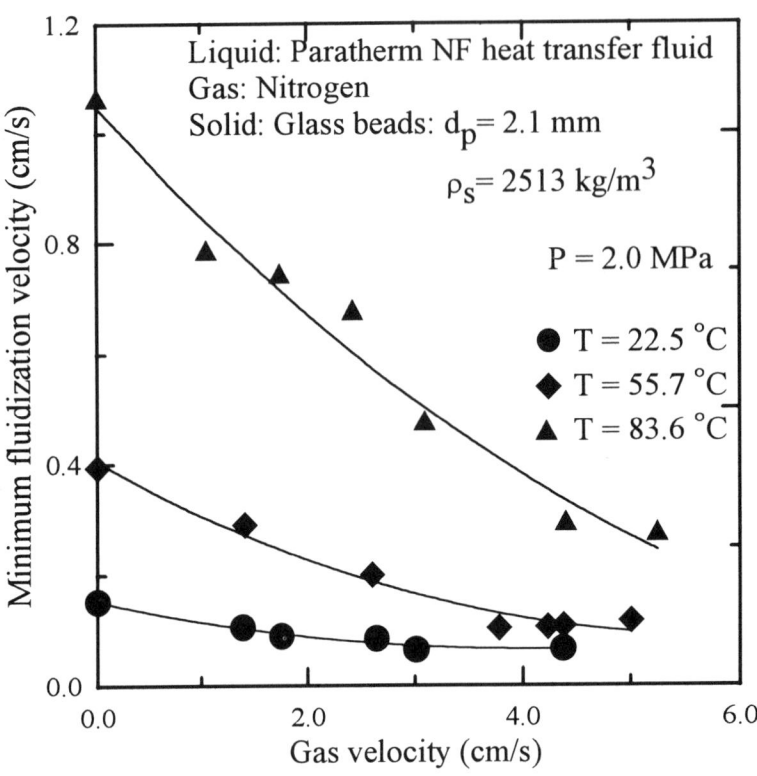

Figure 6.5 Effect of temperature on the minimum fluidization velocity for 2.1 mm glass beads (Jiang et al., 1995b).

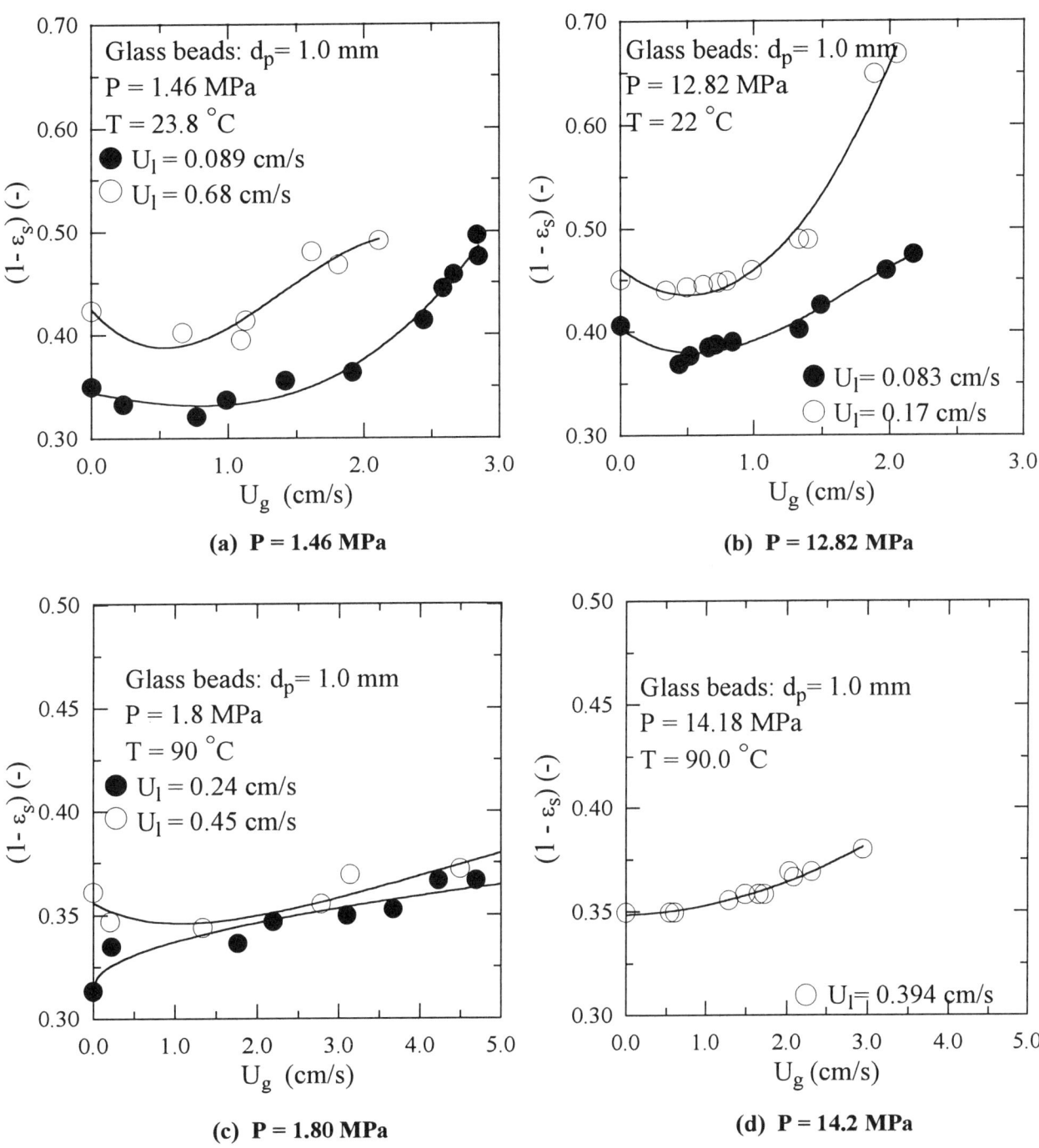

Figure 6.6 Effects of pressure and temperature on the bed expansion and contraction phenomena of a three-phase fluidized bed.

Infinitesimal and Finite Voidage Perturbations in the Compressible-Particle-Phase Description of a Fluidized Bed

P.U. Foscolo, L.G. Gibilaro and S. Rapagnà
Università di L'Aquila, Roio, 67100 L'Aquila, Italy

The Particle Bed Model exploits the analogy of the particle phase of a fluidized bed as a compressible fluid: by drawing on methods well established in the field of ideal gas dynamics, it has been used to analyse both very small and finite perturbations. The linearized equations have been applied extensively to the former case where they deliver a criterion for the stability of the homogeneous state to small disturbances, the minimum bubbling condition, in good agreement with the widely reported findings for both gas and liquid fluidized systems. More recently, attention has focused on the analysis of fully developed, one-dimensional shock-waves that satisfy the non-linear equations of continuity and momentum: for moderate size powders (Geldart Group B), the model predicts a virtually particle free region downstream of a shock front, in accord with the completely void bubble phase manifested by these systems; for finer powders (Geldart Group A), the analysis explains why it is that the bubble phase can contain a small amount of powder, as observed in practice, and also why the well known phenomenon of partial bed collapse, at the homogenous-bubbling transition point, is to be expected.

The mathematical description of a dispersed particle phase as a fluid continuum is implicit, if not explicitly stated, in almost all theoretical treatments of fluidized suspensions. Although clearly convenient, from an analytical point of view, such a notion cannot be said to be immediately acceptable as a credible representation of the physical reality. The usual assumption in the ubiquitous Two-Fluid Model is that both the fluid and the particle phases are to be regarded as *incompressible* continua. On this basis all resemblance to a dispersed particle system is lost.

On the other hand, the notion of the particle phase as a *compressible* continuum accommodates many, clearly visible, features of the dispersed state: a fluidised bed will expand and contract with changing fluid flow rate; it can be compressed, by pushing down the surface with a mesh, and will expand back to its previous state when the mesh is removed; the qualitative similarity to a compressible fluid is self evident.

The quantitative extension of this concept may be readily achieved by considering a particle phase continuum, of density ρ,

$$\rho = \rho_p(1-\varepsilon), \qquad (1)$$

Università di L'Aquila, Roio, 67100 L'Aquila, Italy.

that occupies the entire bed. This is exactly analogous to a true gas with the particles taking the place of molecules. The general equations of change, in one-dimension, for the particle phase of a fluidised suspension,

$$\frac{\partial \rho}{\partial t} + \frac{\partial (\rho u)}{\partial z} = 0, \qquad (2)$$

$$\rho \frac{Du}{Dt} = F - c^2 \frac{\partial \rho}{\partial z}, \qquad (3)$$

then have the exact forms of the continuity and momentum equations for a compressible gas, although the constitutive relationships for the net force acting within the control volume, F, and the dynamic wave velocity, c, are quite different. Under equilibrium conditions, Equation (3) reduces to F = 0, which represents the relationship between ρ and u for a given fluidising velocity.

THE PARTICLE BED MODEL

Constitutive expressions have been derived on the basis of the analogy of the particle phase as a compressible fluid (Foscolo and Gibilaro [1 and 2]). The only empirical input relates to equilibrium expansion data in the expression for F; the dynamic wave velocity, c, was shown to be related to F through:

$$c^2 = \frac{2d}{3} \left.\frac{\partial F}{\partial \rho}\right|_{F=0}. \qquad (4)$$

The force F represents the net effect of fluid-particle interaction and gravity. The interaction force can be decomposed into the drag component (a function of the relative fluid-particle velocity and void fraction) and buoyancy, which represents the effect of the mean pressure gradient in the bed on the particle phase, and is a function of void fraction alone. On this basis, F has been expressed in terms of well established empirical correlations: for the unhindered single particle settling velocity, u_t, and the Richardson-Zaki exponent, n:

$$F = (1-\varepsilon)(\rho_p-\rho_f)g\left[\left(\frac{u_o-u}{u_t}\right)^{\frac{4.8}{n}}\varepsilon^{-3.8} - \varepsilon\right] \quad (5)$$

The dynamic wave velocity, c, is then obtained from Equations (4) and (5):

$$c = \left(3.2gd(1-\varepsilon)(\rho_p-\rho_f)/\rho_p\right)^{0.5} \quad (6)$$

STABILITY ANALYSIS FOR INFINITESIMAL VOIDAGE PERTURBATIONS

The above equations have been employed to predict the stability of the homogeneous, fluidised state to infinitesimal voidage perturbations - in particular the conditions for the onset of bubbling. On linearisation, the velocity variable, in Equations (2) and (3), can be eliminated to yield a single equation for the particle phase density, ρ (or the voidage, ε). This general procedure was first proposed by Wallis [3], who showed that the stability of the homogeneous, fluidised state depended on the relative magnitude of two of the parameters that appear in the linearised equation - the dynamic wave velocity, c, and the kinematic wave velocity, u_ε. An explicit expression for the latter was first derived by Slis et al. [4],

$$u_\varepsilon = (1-\varepsilon) n u_t \varepsilon^{n-1} \quad (7)$$

and also arises from the Particle Bed formulation, which, in addition, provides the dynamic wave velocity, Equation (6).

These expressions lead to the fully predictive form of the Wallis criterion:

$$c - u_\varepsilon = \begin{cases} +ve, & \text{homogeneous fluidisation} \\ 0, & \text{stability limit} \\ -ve, & \text{bubbling fluidisation} \end{cases} \quad (8)$$

Thus, all that is required in order to know whether a bed will be homogeneous or bubbling, at a specified void fraction, are the fluid and particle densities, ρ_f and ρ_p, the particle diameter, d, and the fluid viscosity, μ; from these values u_t and n are evaluated from the standard correlations for use in Equation (8).

This criterion has proved remarkably successful, not only in differentiating between the always-homogeneous and always-bubbling cases (associated with typically liquid and gas fluidised systems respectively), but also in the more delicate task of predicting the voidage at the transition point for those systems that display homogeneous expansion up to a critical fluidising velocity, where $c = u_\varepsilon$, and bubbling behaviour thereafter. This phenomenon has been widely studied experimentally for gas fluidised, fine powders (of, typically, less than 100μm diameter). Figure 1 compares all the available minimum-bubbling voidage data with the predictions of Equation (8); this represents the results of 29 reported investigations [5 to 33].

The results reported in Figure 1 represent systems of wide, as well as narrow, particle size distribution; porous, non-porous, rounded and angular particles; particle materials as diverse as PVC, glass, steel, copper, alumina, hollow fly ash, granular carbon, spent and fresh cracking catalyst; fluids as diverse as air, water and super critical carbon dioxide; gas pressures ranging from 1 to 125 bar; values of ε_{mb} ranging from close to the packed bed value of approximately 0.4 up to nearly 0.85.

Given the range of these reported results, and the fact that they have been obtained using quite different experimental techniques, Figure 1 provides convincing support for the model formulation. A fair degree of experimental scatter is to be expected, particularly with regard to the 'metastable state' that can exist between the minimum-fluidisation and minimum-bubbling conditions [5]: this region, or parts of it, can be stable to infinitesimal perturbations, but unstable to finite ones - a phenomenon that is treated quantitatively in the non-linear analysis reported in the following section; this can lead to premature bubbling, in experiments designed to measure minimum bubbling conditions, as a result of the introduction of measuring instruments, or other obstructions in the bed.

Far fewer systematic data are available for liquid fluidised systems. What there are, however, show good agreement with the criterion - particularly for the case of fine, high density solids (such as copper and lead), that display an abrupt transition to bubbling behaviour, in much the same way found for gas fluidised beds; and the criterion predicts well the effect on the bubbling point of bed temperature (and hence fluid viscosity) and particle size, just as it does for beds fluidised by a gas (Gibilaro et al.[15]): these results are included in Figure 1.

A dimensionless counterpart of Equation (8) has been used to produce a theoretical powder classification diagram, applicable to any powder, fluidised by any fluid (Foscolo et al. [34]); it represents a generalisation of the empirical (Geldart [13]) powder classification, for fluidisation by ambient air. It is reproduced in Figure 2 and will be made use of in the following section; also shown in the figure is the correlation by Grace [35] of extensive gas fluidisation data for the boundary between always-bubbling and transition systems.

STABILITY ANALYSIS FOR FINITE VOIDAGE PERTURBATIONS: THE JUMP CONDITIONS

Numerical integration of the above equations of change, with the constitutive expressions of the Particle Bed Model, Equations (5) and (6), shows that, for systems falling in the always-bubbling or transition regions of Figure 2, a small perturbation develops quickly into a shock wave: that is to say, a discontinuity in voidage, from ε_1 to ε_2, and in velocity, from u_1 to u_2, that travels up the bed with velocity V (Foscolo and Gibilaro [2]). The bubbles observed in practice, under these conditions, are generally very far removed from what could reasonably be regarded as one-dimensional phenomena, thereby casting doubt on the applicability of a one-dimensional model to the study of bubbling behaviour. It turns out, however, that analysis of the full non-linear formulation, in terms of permissible finite voidage perturbations, provides remarkable insight into a number of observed characterstics of the bubbling regime (Brandani and Foscolo [36]).

The basic concepts involved in this analysis are very simple: by considering a coordinate reference system that moves with the shock (at velocity V), transient terms in the governing equations disappear, and neccessary conditions for the exisistance of the shock become, simply, that mass and momentum are conserved across the discontinuity. Once formulated, these conditions (together with the supposition that the shock front simply separates one condition of dynamic equilibrium from another) yield the mathematically permissible jumps in voidage and particle velocity, and the velocity of the front itself.

The condition for conservation of mass across the shock presents no difficulty. For the case of momentum, however, it is not immediately clear how to deal with the constitutive expressions that appear in the conservation equation. This problem is readily resolved by the method proposed by Kotchine [37], and recently adopted, for chemical engineering applications, by Astarita and Ocone [38]. The method involves integrating the terms in the momentum equation over a distance interval that includes the discontinuity at its center, and then evaluating the limiting conditions as the intervals approach zero - thereby retaining solely those terms containing the (infinite) gradient of a variable that experiences a jump.

The 'jump conditions' have been derived, on this basis, for the Particle Bed Model by Brandani and Foscolo [36]. These take the form:

$$[\rho(u-V)] = 0 , \qquad (9)$$

$$\rho_1(u_1-V)[u] - (c_1\rho_p)^2/\rho_1\left[\varepsilon - \frac{\varepsilon^2}{2}\right] = 0, \qquad (10)$$

where the square brackets indicate the changes in value across the jump experienced by the quantities included within them. For the case of dynamic equilibrium (F=0) on both sides of the shock, the above conditions yield both the amplitude, $\Delta\varepsilon$, and the velocity of the discontinuity, V, as a function of ε_1.

In order to illustrate the significance of these results to typical fluidised beds, four examples have been selected that span the full range of behaviour depicted in the generalised powder classification diagram of Figure 2: these examples are indicated in Figure 2 and consist of an always-homogeneous bed, an always-bubbling bed, and two beds that exhibit a transition from the homogeneous to the bubbling state - one close to the always-homogeneous boundary, the other close to the always-bubbling boundary.

Figure 3 shows, for all these cases, the mathematically permissible values of voidage downstream of a shock, ϵ_2, as a function of upstream voidage, ϵ_1. Not all these solutions are physically possible: certainly not those relating to voidages lower than that at minimum fluidisation - assumed here to be equal to 0.4; nor those for which the shock disintegrates due to instability. The criterion for shock stability is given by Wallis [39] and relates to the velocities of the (infinitesimal) kinematic waves on either side of the front: the shock velocity, V, must lie between these two values so that small perturbations on either side run towards the shock-front, instead of giving rise to its degeneration. In Figure 3 all the mathematical solutions are shown, with physically realisable ones distinguished by means of a continuous, rather than a broken, line.

Figure 3a shows always-homogeneous fluidisation to be characterised by the absence of any jump solution, whether physically realisable or not: ϵ_2 remains always equal to ϵ_1. On the other hand, for the always-bubbling case, Figure 3b shows that the only physically realisable solution is ϵ_2 equal to practically unity - corresponding to the virtually particle free bubble phase observed in practice for these systems. It will be noticed, however, that two mathematical solutions appear in the region corresponding to a bed voidage below ϵ_{mf}, the lower one branching upwards from the diagonal at the, physically impossible, minimum bubbling voidage value: $\epsilon_{mb} < \epsilon_{mf}$.

Figures 3c and 3d, that represent systems that show a transition from homogeneous to bubbling fluidisation, illustrate the full significance of the purely mathematical solutions refered to above. In Figure 3c, that corresponds to a transition system close to the always-bubbling boundary, two physically possible solutions appear throughout the region bounded by ϵ_{mf} and ϵ_{mb}, the lower one representing a possible expansion of the homogeneous phase up to the minimum bubbling point, and the upper one (that extends beyond ϵ_{mb}) indicating a bubble phase containing a few percent concentration of solids. This latter feature is in harmony with the reported experimental findings of Grace and Sun [17] for Geldart group A powders. The two posible solutions in the region bounded by the minimum fluidisation and minimum bubbling points provide a theoretical explanation for the 'metastable state' reported for group A systems operated under these conditions: in this region the homogeneous state is stable to infinitesimal perturbations, but unstable to finite ones which can initiate a jump to the upper, bubbling, solution, and, as a consequence, a collapse of the homogeneous phase, perhaps right back to the minimum fluidisation voidage.

Figure 3d, which represents a transition-region system close, this time, to the always-homogeneous boundary (that can therefore be thought to relate particularly to liquid fluidised beds), shows that the metastable region, where two physically realisable solutions are to be found, is limited to the vicinity of the minimum bubbling point; this provides an explanation for the limited extent of bed collapse observed for gas fluidised beds operated under high pressure conditions (Jacob and Weimer [19]), and the complete absence of dense phase contraction reported for liquid fluidised systems (Gibilaro et al. [15]).

CONCLUSIONS

The analogy of the particle phase of a fluidised bed as an ideal, compressible fluid leads to a predictive account of dynamic behaviour. For small perturbations, the linearised equations yield estimates of the minimum bubbling voidage in good agreement with the copious experimental evidence. For large perturbations, the full non-linear equations provide insight into the phenomenon of bubbling fluidisation: in particular, they offer an explanation for the metastable expansion regime observed between minimum fluidisation and minimum bubbling conditions, and yield estimates for the possible extent of dense phase contraction that is a consequence of this phenomenon; they also reveal the conditions under which the bubble phase can be expected to contain a small concentration of particles.

ACKNOWLEDGEMENT

This work was funded by a grant from the

Italian Consiglio Nazionale delle Ricerche (Tecnologie Chimiche Innovative).

NOTATION

Ar	Archimedes number, $gd^3\rho_f(\rho_p-\rho_f)/\mu^2$	
c	dynamic wave velocity	m/s
d	particle diameter	m
De	density number, ρ_f/ρ_p	
F	net force on particle phase	N/m^3
g	gravitational field strength	N/kg
n	Richardson-Zaki parameter	
t	time	s
u	particle phase velocity	m/s
u_o	fluidising velocity	m/s
u_t	unhindered particle velocity	m/s
u_ε	kinematic wave velocity	m/s
z	distance	m
ε	void fraction	
ε_{mb}	minimum bubbling voidage	
ε_{mf}	minimum fluidisation voidage	
μ	fluid viscosity	Ns/m^2
ρ	particle phase density	kg/m^3
ρ_f	fluid density	kg/m^3
ρ_p	particle density	kg/m^3

LITERATURE CITED

1. Foscolo, P.U. and L.G. Gibilaro, *Chem. Engng Sci.*, **39**, 1667 (1984).

2. Foscolo, P.U. and L.G. Gibilaro, *Chem. Engng Sci.*, **42**, 1489 (1987).

3. Wallis, G.B., "One-dimensional Waves in Two-component Flow", UKAEA Report AEEW-R162 (1962).

4. Slis, P.L., Th.W. Willemse and H. Kramers, *Appl. Sci. Res.*, **A8**, 209 (1959)

5. Abrahamsen, A. R. and D. Geldart, *Powder Technol.*, **26**, 35 (1980).

6. Abrahamsen, A.R. and D. Geldart, *Powder Technol.*, **26**, 47 (1980).

7. Barreto, G. F., "Behaviour of Beds of Fine Powders Fluidised by Gases at Pressures of up to 20 Bar", PhD Thesis, Univ. of London, (1984).

8. Crowther, M. E. and J. C. Whitehead, "Fluidization of Fine Particle at Elevated Pressure," in *Fluidization*, J. F. Davidson and D. L. Keairns, (Eds.), Cambridge University Press, Cambridge, 65 (1978).

9. De Jong, J. A. H. and J. F. Nomden, *Powder Technol.*, **9**, 91 (1974).

10. Dry, R. J., M. R. Judd and T. Shingles, *Powder Technol.*, **34**, 213 (1983).

11. Foscolo, P. U., R. Di Felice and L. G. Gibilaro, *Chem. Eng. Process.*, **22**, 69 (1987).

12. Foscolo, P. U., A. Germanà, R. Di Felice and L. De Luca, "An Experimental Study of the Expansion Characteristics of Fluidized Beds of Fine Catalysts Under Pressure," in *Fluidization VI*, J. R. Grace, L. W. Shemilt and M. A. Bergougnou, (Eds.), Engineering Foundation, New York, 187 (1989).

13. Geldart, D., *Powder Technol.*, **7**, 285 (1973).

14. Geldart, D. and A. C. Y. Wong, *Chem. Engng Sci.*, **39**, 1481 (1984).

15. Gibilaro, L. G., I. Hossain and P. U. Foscolo, *Can. J. of Chem. Eng.*, **64**, 931 (1986).

16. Godard, K. and J. F. Richardson, *Inst. Chem. Eng. Symp. Ser.*, **30**, 126 (1968).

17. Grace, J. R. and G. Sun, *Can. J. of Chem. Eng.*, **69**, 1126 (1991).

18. Guedes de Carvalho, J. R. F., *Chem. Engng Sci.*, **36**, 413 (1981).

19. Jacob, K. V. and A. W. Weimer, *AIChE J.*, **34**, 1395 (1987).

20. Khoe, G. K., T. L. Ip and J. R. Grace, *Powder Technol.*, **66**, 127 (1991).

21. King, D. F. and D. Harrison, *Trans. Inst. Chem. Engrs.*, **60**, 26 (1982).

22. Kono, H. O., S. Chiba, T. Ells, M. Daniell and M. Suzuki, *World Congress III of Chemical Engineering*, Paper 81-108, Tokyo, (1986).

23. Massimilla, L., G. Donsì, and C. Zucchini, *Chem. Engng Sci.*, **27**, 2005 (1972).

24. Mutsers, S. M. P. and K. Rietema, *Powder Technol.*, **18**, 239 (1977).

25. Piepers, H. W., E. J. E. Cottaar, A. H. M. Verkooijen and K. Rietema, *Powder Technol.*, **37**, 55 (1984).

26. Piepers, H. W. and K. Rietema, "Effects of Pressure and Type of Gas on Gas-Solid Fluidization Behaviour," in *Fluidization VI*, J. R. Grace, L.W. Shemilt and M. A. Bergougnou, (Eds.), Engineering Foundation, New York, 203, (1989).

27. Poletto, M., "Stability of Particle Beds Fluidised by CO_2 at Supercritical Conditions", PhD Thesis, Univ. of Naples, (1992).

28. Rapagnà, S., R. Di Felice, P. U. Foscolo and L. G. Gibilaro, "Experimental Verification of the Scaling Rules for Fine Powder Fluidization," in *Fluidization VII*, O. E. Potter and D. J. Nicklin, (Eds.), Engineering Foundation, New York, 579 (1992).

29. Rapagnà, S., P. U. Foscolo and L. G. Gibilaro, *Int. J. Multiphase Flow*, **20**, 305 (1994).

30. Rietema, K. and S. M. P. Mutsers, in *Fluidization*, Cambridge University Press, Cambridge, (1978).

31. Rowe, P. N., *AIChE Annual Meeting*, Miami Beach, paper No. 58f (1986).

32. Simone, S. and P. Harriott, *Powder Technol.*, **26**, 161 (1980).

33. Sobreiro, L. E. L. and J. L. F. Monteiro, *Powder Technol.*, **33**, 95 (1982).

34. Foscolo, P.U., L.G. Gibilaro and R. Di Felice, *Appl. Sci. Res.*, **48**, 315 (1991).

35. Grace, J.R., *Can. J. of Chem. Eng.*, **64**, 353 (1986).

36. Brandani, S. and P.U. Foscolo, *Chem. Engng Sci.*, **49**, 611 (1994).

37. Kotchine, N.E., "Sur la Theorie des Ondes de Choc dans un Fluide", *Circ. Mat. Palermo*, **50**, 305 (1926).

38. Astarita, G. and R. Ocone, "Discontinuities and Interfacies in Transport Phenomena Coupled with Kinetics and Relaxation", in *Advanced Transport Processes*, Elsevier, Amsterdam, 319 (1992).

39. Wallis, G.B., "One-dimensional Two-phase Flow", McGraw Hill, New York (1969).

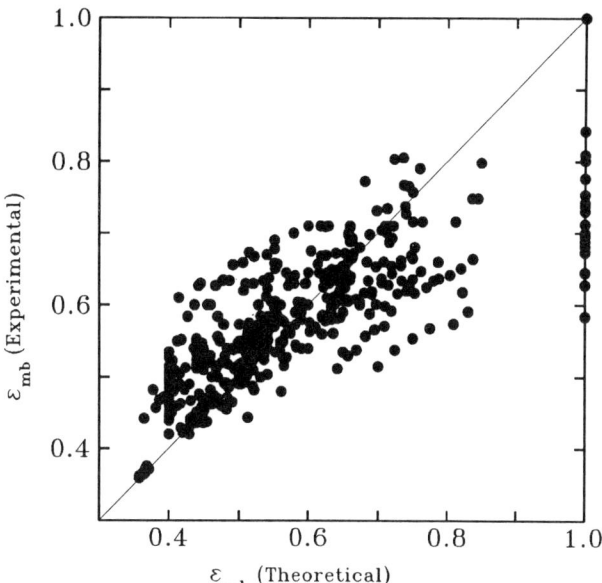

FIGURE 1: Comparison of all available experimental observations of ε_{mb} with predictions of the Particle Bed Model (references given text).

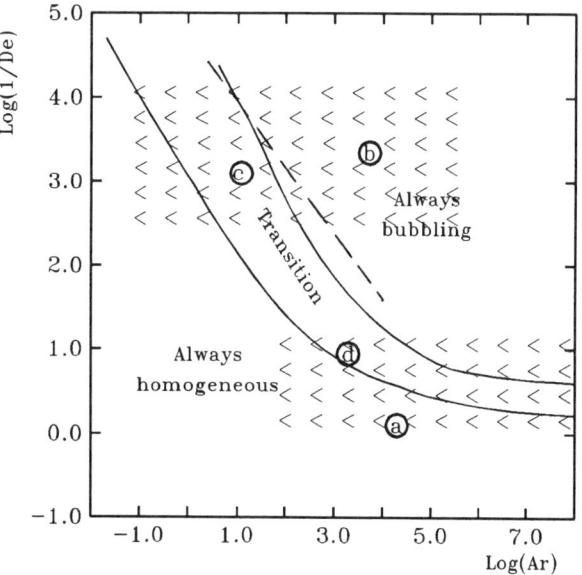

FIGURE 2: Theoretical classification map for the fluidisation of any powder by any fluid. The upper and lower shaded areas refer, respectively, to typical gas and water fluidised systems. The empirical correlation of Grace (1986) for the transition / always-bubbling boundary is shown as a broken line. The labelled points relate to the specific systems of Figure 3.

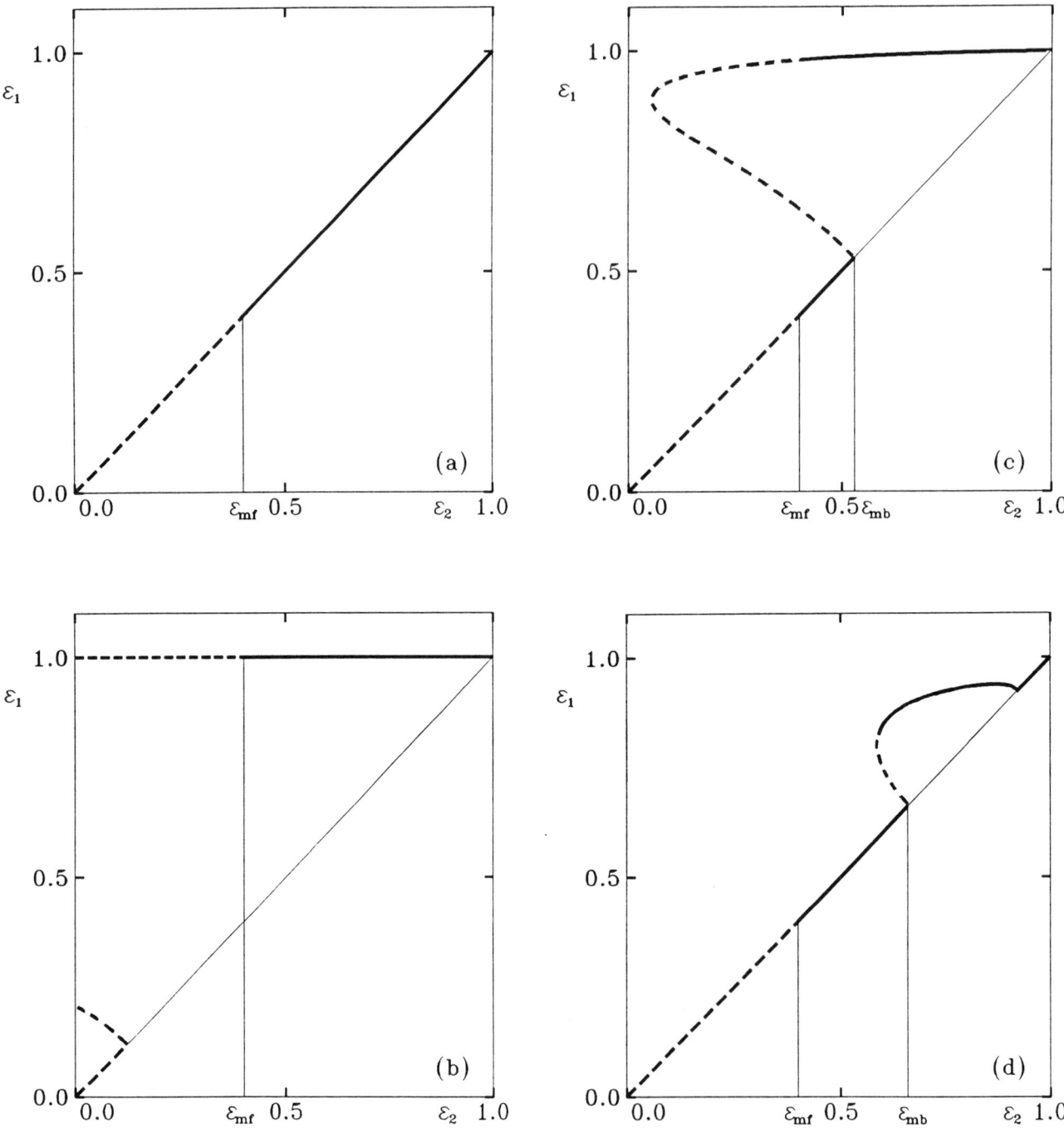

FIGURE 3: Solutions obtained from the jump conditions, Equations (9) and (10), for the systems labelled in Figure 2: a – always-homogeneous; b – always-bubbling, c – transition system, near the always-bubbling boundary; d – transition system, near the always-homogeneous boundary.

Investigation of Dispersion Characteristics of a Fluidized Bed Reactor

C. Ercan, S.C. Arnold and H.E. Barner
ABB Lummus Global Inc., 1515 Broad Street, Bloomfield, NJ 07003

Dispersion characteristics of a fluidized bed reactor were investigated in a large, three-dimensional cold-flow model with rectangular cross section of 0.46m by 0.61m which is a size large enough to largely eliminate any substantial wall effect. The cold model was designed to simulate an actual commercial reactor configuration with deep bed, internals, freeboard disengagement, cyclones and diplegs. Axial and radial dispersion characteristics of the system were characterized by dispersion coefficients. A steady-state tracer injection technique was used. A known amount of an inert tracer was injected into the bed, and the concentration of the tracer was determined at various axial and radial positions. From these concentration profiles, axial and radial dispersion coefficients were determined. Effects of internals and superficial velocity on axial and radial dispersion were determined. Superficial gas velocity range was wide enough to cover a range from bubbling to turbulent regimes. Superficial velocity and internals have substantial effect on axial dispersion at low superficial velocity (bubbling bed regime) while their effects diminish as the turbulent regime is approached. Radial dispersion is affected by superficial velocity, but is relatively insensitive to internals. Details of these findings and their implications for commercial design are discussed.

It is generally desirable to approach plug flow in chemical reactors in order to maximize product yield. Low gas axial dispersion and high gas radial dispersion are thus favorable hydrodynamic characteristics for a fluid bed reactor. Baffles and other types of internals are sometimes employed in order to improve reactor performance by controlling backmixing. Catalyst particle size distribution and the presence of sufficiently fine particles are also recognized to be of importance in promoting favorable hydrodynamics.

Summaries of dispersion characteristics of fluid bed reactors have been reported by Avidan [1] and Grace [2]. Although some studies have been made on gas axial dispersion, they are mainly limited to the bubbling bed regime in relatively small-scale units. Radial gas mixing studies are limited. The present study was therefore undertaken to provide further insight on both gas axial and radial mixing in a large, three-dimensional cold flow model. Test conditions covered a range from the bubbling to the turbulent regime, and the effects of various reactor internals and particle size distributions were investigated.

It was found that axial gas dispersion ("backmixing") is reduced by increasing superficial gas velocity and also by the inclusion of baffles. Radial mixing also shows a tendency to decrease with increasing superficial gas velocity. Implications for design depend somewhat on the specific application and reactor size (bed diameter and bed depth). Generally, there is greater concern in minimizing axial gas dispersion (relative to radial dispersion), supporting the use of a relatively high gas velocity (turbulent regime) and selective use of internal baffles.

EXPERIMENTAL

The test column, made of plexiglass, had a rectangular cross section of 0.46 by 0.61m and an overall height of 8.2m (Figure 1). Air entered the column through inlet nozzles in a distributor grid at the bottom and flowed upward, fluidizing catalyst particles in the column. A small amount of additional air entered through a supplemental gas sparger at an elevation of approximately 0.3m.

Two sets of simulated heat transfer coil banks were placed between elevations of 1.30 and 1.75m, and 2.74 and 3.2m, respectively. Provisions were made to install removal horizontal baffles approximately 0.5m above the top of each coil bank (at elevations of 2.24 and 3.68m). They could also be lowered to 0.5m above the top of each coil bank, i.e., to elevations of 1.80 and 3.25 m. This paper reports results of a representative group of baffle arrangements which were tested, along with tests without baffles.

A commercialized fluidized-bed catalyst was used in all tests. Catalyst with two particle size distributions was used. "A-type" catalyst represents a relatively coarse, narrow particle size range, whereas "B-type" represents a smaller average particle size and wider size distribution than A-type. (These designations should not be confused with Geldart "A" vs. "B" classifications. Both the A-type and B-type catalysts used in this study had average particle sizes smaller than 100 micrometers and were within the Geldart A classification range). All entrained catalyst was recovered from the offgas by cyclones and returned to the column through diplegs. Superficial gas velocity, as measured at the top of the bed, was varied between 0.3 and 0.75m/s. This range is considered to cover the regimes from bubbling to turbulent.

A steady-state experimental tracer technique was used, consisting of continuous injection of a known amount of helium at a given height, and detecting the concentration of helium at a different elevation upstream or downstream of the injection point. Tests focussed on the investigation of the effects of superficial velocity, baffle configuration and catalyst particle size distribution on three hydrodynamic characteristics of the bed: axial density profile, gas axial and radial mixing. This paper is limited to the results on mixing characteristics.

The gas axial dispersion coefficient, D_a, was determined from the upstream measurement of the tracer concentration profile. It was calculated from the one-dimensional diffusion equation, i.e., from the slope of the tracer concentration curve from injection to detection points. The gas radial dispersion coefficient, D_r, was determined from the downstream measurement of the tracer concentration profile. This coefficient was calculated from Klinder's diffusion equation, a relatively complex equation, whose solution required both axial and radial dispersion coefficients to predict the radial profile. Here, the experimentally measured axial dispersion coefficient and trial-and-error guessed radial dispersion coefficient were used to match a predicted radial concentration profile with the experimentally measured one.

RESULTS

Typical experimental results for axial and radial gas dispersion are presented below. The test series and baffle configurations are summarized in Table 1. The actual body of data that has generated is much greater than can be included in this paper. Overall assessments and conclusions, however, are based on the full body of information that has been developed.

GAS BACKMIXING

Typical results for the axial dispersion coefficient measured in the absence of baffles are plotted in Figure 2 as a function of bed elevation. The coil bank reduced axial dispersion at the lowest superficial velocity (in the bubbling bed regime), but did not significantly impact axial dispersion at the higher superficial velocities.

Figure 3 shows a similar plot for a bed having both coil banks and baffles. At all superficial velocities, the axial dispersion coefficient across the baffles is reduced to approximately half the value for open bed dispersion at high superficial velocity. This demonstrates that baffles are effective tools for reducing backmixing, with the greatest benefit at low velocity. Comparison of Figures 2 and 3 indicates that at low gas velocity backmixing across the lower coil bank is greater for runs with baffles than for runs without baffles. This is believed to be caused by the presence of the baffle above the coil bank; this tendency was generally observed in all tests with various baffle configurations.

Figures 4-6 depict the impact of superficial velocity on the dispersion coefficient for various bed internals and catalyst particle size distribution. These figures are for different vertical sections of the bed. The dispersion measurements in Figure 4 were for injection at 1.9m and detection at 1.1m, covering the region of the first coil bank. For Run A3, the region also included the first baffle, located in this case only 0.05m above the first coil bank. A general characteristic of all the curves across the first coil bank is that the dispersion coefficient is high at low velocity, decreases with increasing superficial gas velocity and then levels off. Differences between the curves are the results of different downstream baffle arrangements and catalyst particle size distributions.

The curve for Run A3 in Figure 4 represents the effectiveness of the baffle. This curve appears to show an upward trend with increasing gas superficial velocity.

This does not represent a true trend, but is an artifact of the calculation based on the same assumed value of C/C_∞ of 0.005 for all these data points, for which the measured concentration levels were too low to determine accurately. Assumption of a constant concentration at varying velocity naturally given an increasing trend. The plotted dispersion coefficient values, however, represent the upper limit for the A3 tests.

Dispersion coefficients for the next region, with tracer injected at 2.6m and detected at 2.0m, are shown in Figure 5. This region included the first baffle for some of the tests and no baffle for others. Points applicable to cross the baffles, and points applicable to the open region between baffles and coil bank, can be easily distinguished. The observed trends are similar to those discussed for Figure 4: dispersion decreases with velocity across the open region, while dispersion is more or less constant across the baffle.

Figure 6 displays dispersion coefficients obtained in the third measured region: injection at 3.4m and detection at 2.5m This region contained the second coil bank, and in the case of Run A3, also the second baffle. Again, the same general trends are observed. D_a across the coil bank decreases modestly with increasing gas superficial velocity. However, at the low velocity end, the dispersion coefficient across the second coil bank is less than the value across the first coil bank (Figure 4). This is attributed to a lower bed height above the injection point, resulting in a reduction of the backmixing effect imposed by the baffle above it.

Careful analyses of all the figures reveals that the baffle type and particle size distribution (the **A** runs vs. **B** runs) might have small effects on gas axial dispersion, perhaps more pronounced in the low velocity region. These effects, however, seem small relative to the scatter of the data.

The general trend for axial gas backmixing observed here is consistent with that reported by Cankurt and Yerushalmi [3] in a small diameter unit. They calculated that the extent of gas backmixing diminishes in the turbulent regime and that plug flow is approached in the fast fluidization regime.

RADIAL MIXING

The effect of various baffles on radial concentration profiles is shown in Figures 7 and 8 at two different gas superficial velocities. In both cases, baffles were not found to have significant effects. In some cases, off-centered and one-side-biased curves are seen; this reflects some horizontal movement of the upward gas flow within the fluidized bed column. In general, the magnitude of concentration variations is greater at high gas velocity (Figure 8) than at low gas velocity (Figure 7).

Radial dispersion coefficients for a typical case with baffles are plotted in Figure 9 as a function of superficial gas velocity. It was difficult to find differences between the different regions across coil banks or baffles. This was also true for the open regions between coil banks for tests without baffles. On the other hand, a consistent decreasing trend is observed as superficial velocity is increased. Review of the data for the different particle size distributions does not reveal any impact.

CONCLUSIONS

Axial gas dispersion coefficients decrease with increasing superficial gas velocity and level off in the turbulent fluidization regime. The presence of heat transfer tube banks in the bed has some effect, but only a low gas velocity.

Axial gas dispersion coefficients across baffles are about one-half the limiting values in a baffle-free bed; baffles are therefore effective in reducing backmixing, especially at low gas velocities (bubbling regime).

Radial gas dispersion coefficients also decrease with increasing superficial gas velocity. The impact of baffles appears to be minor.

Baffle type and catalyst particle size distribution in the ranges investigated have negligible effects on both axial and radial dispersion.

IMPLICATIONS FOR DESIGN

This study indicates that (undesirable) backmixing in a fluidized bed reactor can be minimized by operating in the turbulent (as opposed to bubbling) fluidization regime.

Furthermore, the use of horizontal baffles can be useful in further reducing backmixing. On the other hand, high gas velocities (turbulent regime) tend to reduce radial mixing (which is undesirable); distribution of the reactants uniformly across the reactor cross section is thus very important to minimize creation of radial concentration gradients.

The design equations for a specific reactor geometry and reaction kinetics need to be solved in order to develop more quantitative interpretations of the dispersion characteristics. In such an analysis, both dispersion coefficients enter into the calculation as dimensionless Peclet numbers which are functions of velocities and reactor dimensions, as well Such calculations are useful to estimate the impact of different levels of the dispersion coefficients, and to determine, for example, the effects of gas velocity and bed height and diameter, and their interactions. Quantification of these results is beyond the scope of the present paper.

LITERATURE CITED

1. Avidan, A., "Turbulent Fluid Bed Reactors Using Fine Powder Catalysts," presented at Joint Meetings of Chemical Engineering, AIChE-CIESC, Beijing (1982).

2. Grace, J.R., *Chem. Eng. Sci.*, **4-5**, 1953 (1990).

3. Cankurt, N.T. and J. Yerushalmi, "Gas Backmixing in High Velocity Fluidized Bed" in *Fluidization*, Davidson, J.F. and D.L. Keairns, (Ed.), Cambridge Univ. Press (1948).

Table 1.   CATALYST TYPES AND BAFFLE CONFIGURATIONS TESTED

Test Series	Catalyst Type	Baffle Configuration
A0	A	No Baffle.
A1	A	1. Perforated plate, with base open area, placed at 224 cm elevation. 2. Perforated plate, with base open area, placed at 368 cm elevation.
A2	A	1. Perforated plate, with 1.9 times base open area, placed at 224 cm elevation. 2. Perforated plate with 1.6 times base open area, placed at 368 cm elevation.
A3	A	1. Perforated plate, with 1.9 times base open area, placed at 180 cm elevation. 2. Perforated plate, with 1.6 times base open area, placed at 325 cm elevation.
A4	A	1. Proprietary vendor baffle, with base open area, placed at 236 cm elevation. 2. Proprietary vendor baffle, with 1.5 times base open area, placed at 381 cm elevation.
B0	B	No Baffle.
B1	B	1. Perforated plate, with base open area, placed at 224 cm elevation. 2. Proprietary vendor baffle, with 1.6 times base open area, placed at 368 cm elevation.

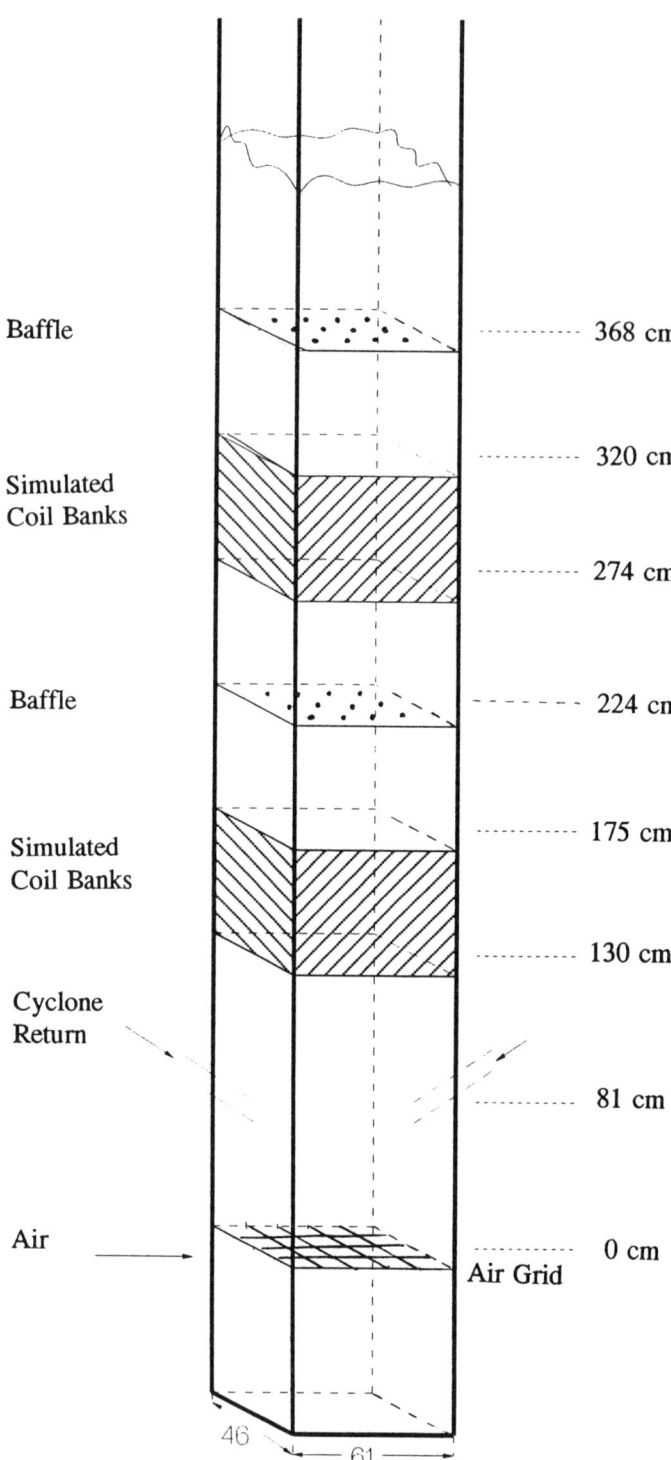

Figure 1. Schematic View of Cold Flow Unit

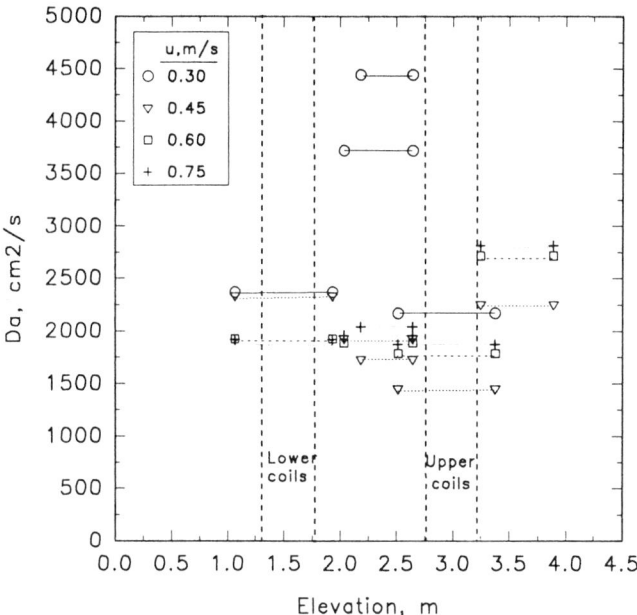

Figure 2. Average axial dispersion values at various velocities measured along the reactor, for B0 tests

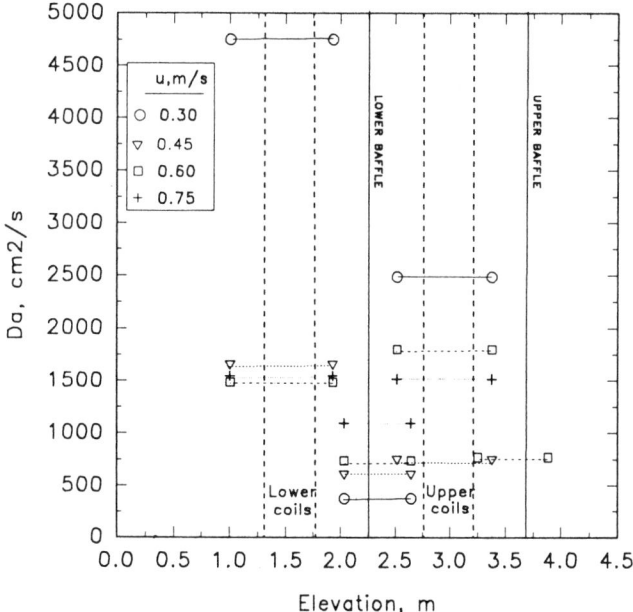

Figure 3. Average axial dispersion values at various velocities measured along the reactor, for A1 tests

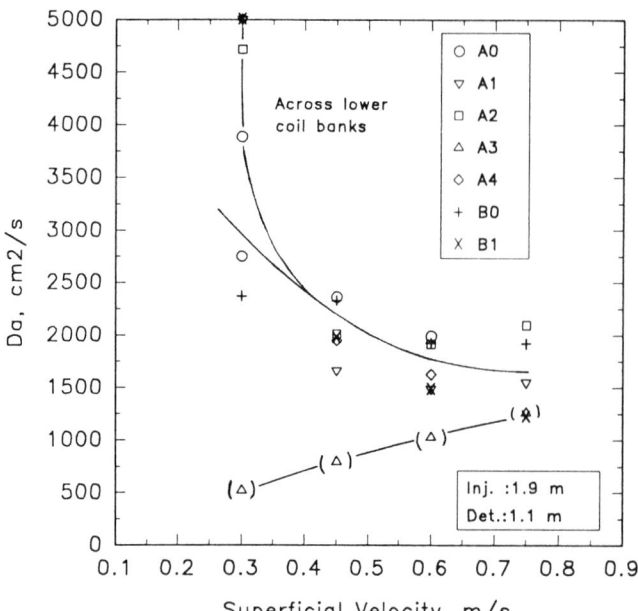

Figure 4. Effect of various internals at superficial velocity on gas axial dispersion coefficient
( ) designates Da based on $C/C_\infty = 0.005$

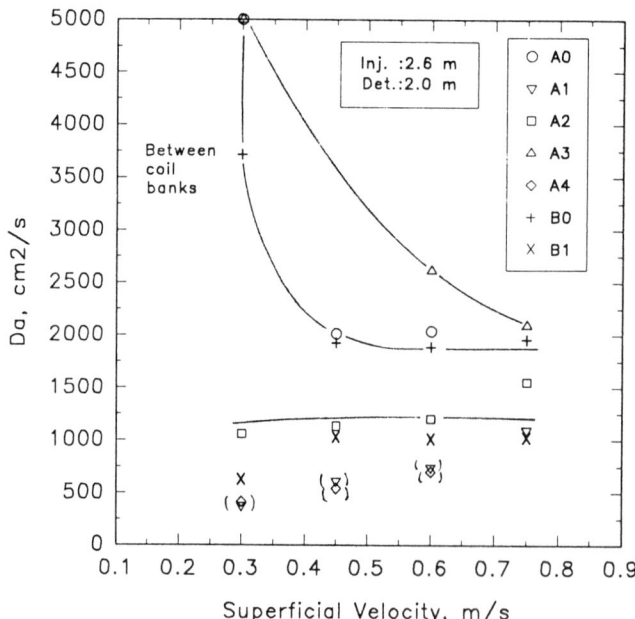

Figure 5. Effect of various internals and superficial velocity on gas axial dispersion coefficient
( ) designates Da based on $C/C_\infty = 0.005$

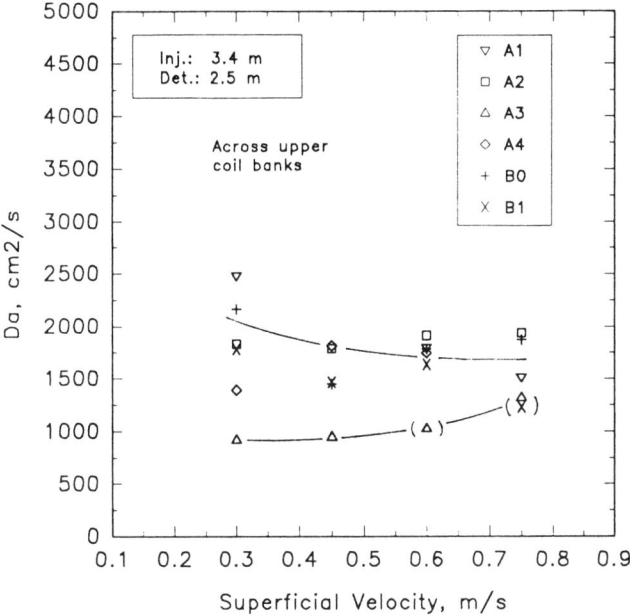

Figure 6. Effect of various internals and superficial velocity on gas axial dispersion coefficient
( ) designates Da based on $C/C_\infty = 0.005$

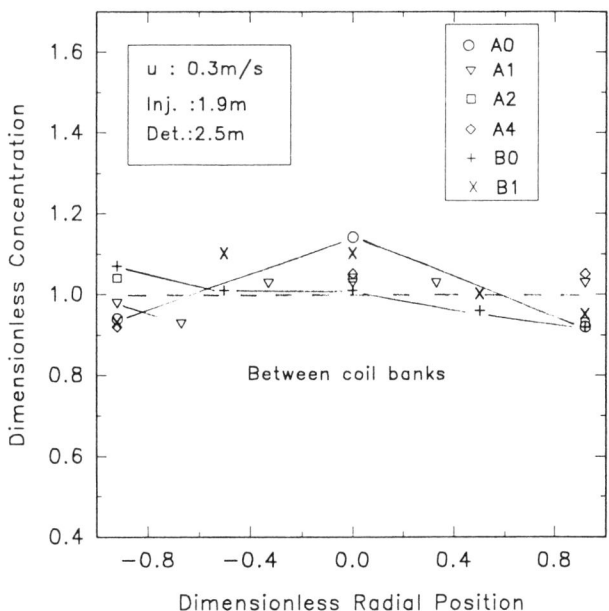

Figure 7. Effect of various parameters on radial concentration profile

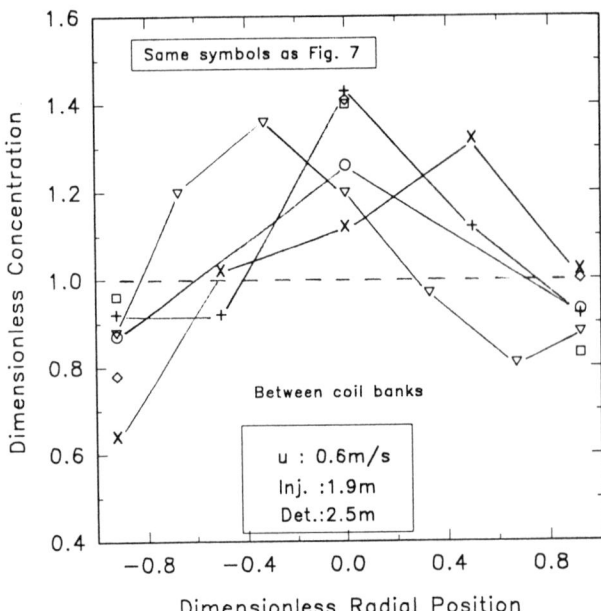

Figure 8. Effect of various parameters on radial concentration profile

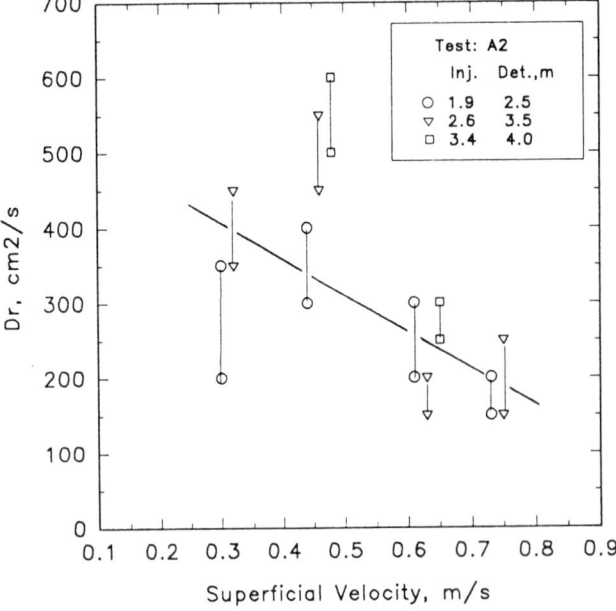

Figure 9. Effect of various parameters on radial dispersion coefficient

Basic Approach on the Control Systems for Granulation Processes: Case of Fluidized Bed Granulators

H.O. Kono and J.J. Su

Department of Chemical Engineering, West Virginia University,
Morgantown, WV 26506

A novel granulator's control concept and approach were introduced in this work for the continuous mode production of small granules (0.5 to 3 mm) in a fully automated granulation pilot plant, consisting of the fluidized bed granulator, dryer, product granule screen, seed granule generating unit etc. (100 to 250 kg/hr). First, an appropriate granulation condition was found under which the function of granule's seed nuclei generation and growth of granules can independently be controlled. Secondly, this control concept was verified in the pilot plant and 72 hours continuous automatic operation had been carried out. Thirdly, a mathematical model was developed and compared with the experimental data.

The basic approach to control the granulation process for the production of small granules (0.5-3 mm) was investigated in a complete set of granulation pilot plant unit (100-250 kg/hr), using fine ceramic powders (10-150 μ) as feed stock and an aqueous liquid as binder. This pilot plant consisted of not only a granulator but also all the related component units such as screening, drying, seed recycling, seed grinding, and other controlling units. The total system was fully automated and a steady state continuous mode operation was accomplished.

To control the continuous mode operation of granulators, the following two key factors should carefully be taken into consideration, which unfortunately have not become aware of by most of the papers on granulation mechanisms reported in the past literature. The first one is well known granule's growth mechanism and the other is the generation mechanism of granule's nuclei. When the granulation process is to be controlled simply by the mixing ratio of feed powder and liquid binder, the results will generally be disastrous. However, the recognition of this fact has been overlooked frequently in the past mathematical models. To improve this situation, a novel seed control procedure was proposed here and experimentally verified through a steady state operation of pilot plant systems.

It should be the purpose of granulation process to simultaneously satisfy the following requirements:

(a) Making a certain size of granules from fine powders (enlargement)

(b) Having a certain voidage of granules to attain a specific mechanical strength (densification)

Particularly, the recent market requires generally the product granules to be strong enough to be easily processed and weak enough to be easily dispersed when necessary. Therefore, granulation processes should be developed to satisfy the above requirement at the same time. Accordingly, the granulation mechanism is also to be investigated from the identical view point. Thus it is rather obvious that the classical approach to granulation mechanism and related population balance model are to be reconsidered fundamentally again.

In this study, we focussed on several operation parameters of the granulation process. Firstly, the product granule size is to be controlled and determined by controlling the size and numbers of the seed recycle. Secondly, the growth mechanism is to controlled by selecting and maintaining an appropriate voidage saturation fraction [1,2,3], S (the fraction of the voidage in green granule filled by liquid binder). Thirdly, the distribution of S is to be controlled by mixing condition of grown-up granules (α), intermediate granules (β) and prewetted powder (γ) in the granulator [4,5,6]. Fourthly, the mean residence time (t^*) is to be controlled by the held-up amount of granule materials in the granulator. Fifthly, the solid particles force prevailing in the granulators. Finally, all the other related component units (screening, drying, etc.) are to be operated carefully, preventing any additional effects on the granulation process [7,8].

With respect to fundamental granulation mechanism, the basic concepts of S, S* and distribution of S in the granulation should be explained here. The saturation fraction of S was originally defined by Rumpf in his pioneering article in 1958 [9] to describe the mechanical structure of a green granule. Kono [1] extended this concept to express the granulation mechanism in pan granulators. In stead of using liquid/powder ratio, S* was introduced to represent the characteristic granulation mechanism [2,3]. The modified liquid saturation fraction S* means a mean value of S of granule and other intermediate materials held in the granulators. The liquid/powder ratio of granules and its intermediate material in granulators is not homogeneous but there is a characteristic distribution. The S-distribution was introduced to express the above heterogeneity. Thus, all the parameters of S, S* and S-distribution are crucial to characterize the granulation mechanism [6].

Based upon the approach proposed, the 72 hours continuous operations were carried out, using the fully automated granulation pilot plant. The significance of this work is to verify the proposed granulation mechanism and the performance of the total granulation system experimentally. Through a series of operations, the proposed granulation mechanism seems to be in agreement with operation results and fully automated 72 hours continuous operation of the pilot plant had been accomplished. The pilot plant performance together with the granulation mechanism and model described in this work is investigated for the case of system using fluidized bed granulator (FBG). Nevertheless, the principles of granulating processes seem to be pretty in common regardless of types of granulators.

THEORETICAL BACKGROUND

Selection of Appropriate Granulating Mechanism

The general granulation mechanism can be schematically shown in Figure 1. As granulation mechanisms [3], there are two granulation patterns; e.g., when or where $S^*=0.7$ to 0.8, layering granulation (Type α and/or β granulation) occurs and when $S^*=0.3$ to 0.7, collision granulation (Type γ) granulation occurs. In the layering granulation, more or less dry powders adhere directly to the wet surfaces of granules. The granule coated by dry powder will get wet and renew its surface. In the collision granulation, the feed powder first gets wet forming at the same time minute, soft granules. Then granules grow in size by the collision (γ granulation) mechanism, generally generating a large number of granule seed nuclei [6].

These two different granulation mechanisms take place simultaneously in any granulators as shown in Figure 1. However it is important to distinguish the difference of these two granulation mechanisms. When the γ type granulation occurs more, the generation of seeds can also occur more. Then, the granule growth and the granule seed generation will take place side by side, inducing the generation of too many seed nuclei, making the granulation process out of control [4].

To quantitatively express the occurrence of γ-granulation in granulators, γ-value (weight fraction of prewetted powder against whole hold-up material in granulator) is defined, which should generally be kept smaller than 0.10 for smooth operation. Within the range of operating conditions of $S^*=0.7$ to 0.8 and $t^*=0.33$ to 0.66 hr.

For the granulation of the ceramic powders with an aqueous binder used in this study, the following operating conditions were experimentally found to be appropriate through preliminary test; i.e., $S^*=0.7$ to 0.8 (funicular stage), and granule's mean residence time$=0.33$ to 0.66 hour, under which the γ-value could be controlled with γ-value smaller than 0.1. Note that these characteristics are to be determined by the properties of powders and binder liquid. Under the specified condition, the sample powders could be kept at a stable α-β type granulation mechanism, and the green product granule's size and its distribution are to be controlled by simply adjusting the number and size of seed nuclei.

Estimation of Product Granules' Size Distribution under the Seed Control Procedure Condition

Selecting an appropriating granulation condition, a granulation mechanism can intentionally be realized, under which only the granule growth are occurring smoothly without any granule's seed generation. The designed seed granules of the appropriate size and numbers are to be fed into the system with a designed feed rate. So that the operation of granulators can be scientifically designed by the seed control. In this work, an approach is proposed, using of seeds with an appropriate size and amount. The population balance of granules should be independently accomplished using seeds, while maintaining S^* and t^* at an appropriate values. Note that the quality and amount of seed granules can be generated in the total production system, using appropriate screen and grinding equipments.

When the layering granulation is predominantly prevailing, the growth rate of a spherical granule in an FBG can approximately be written as Equation (1)

$$\frac{d(\frac{\pi}{6}\rho_p D^3)}{dt} = k'\pi D^2 \quad (1)$$

where

$$k' = c_1 S^* \quad (2)$$

k' is the granulation rate coefficient, assumed to be proportional to modified liquid binder saturation fraction, S^*. The growth rate of granules in the FBG is expressed as Equation (3)

$$\frac{dD}{dt} = k'\frac{2}{\rho_p} \quad (3)$$

$$k = \frac{2k'}{\rho_p}$$

Supposing the liquid content of granules is maintained at an optimum value, the size distribution of product granules out of the FBG can be mathematically expressed. Figure 2 shows a rough material balance.

The following assumptions are made:

(a) The layering granulation mechanism is predominantly prevailing [1].
(b) The granulation process is maintained at a steady state condition.
(c) The optimum liquid-powder feed ratio is to be chosen so as to attain the optimum S^* [2].
(d) The seed generation rate is to be adjusted by feeding a certain number of seeds from the outside of the granulator. In other words, the S^* is chosen to suppress the generation of seed inside the granulator. The appropriate selection of S^* is also very important to avoid "granule-granule" formation, called sometimes as wild granulation [6].
(e) The number of the product granules are ideally to be the same as the number of seeds fed in, if seed-effectiveness factor is unity. However, the reality was not like that. As shown in Table III, the effective seed size can actually be determined by experiments. The size of granule's seed depends upon the specific granulation condition. Nevertheless, these necessary data can easily and quickly be obtained through lab-scale test.

(f) A well mixing of granules is to be attained in the granulators.
(g) Elutriated small granules are collected by a cyclone and returned back to the bed.
(h) The attrition can be adjusted as almost negligible by the appropriate selection of S^*, e.g., $S^* = 0.7$ to 0.8 [5].

The size distribution functions of seeds and granules, and the residence time distribution are defined as follows.

$$\int_{D_{s,min}}^{D_{s,max}} f_s(D_s) dD_s = 1 \quad (4)$$

$$\int_{D_{min}}^{D_{max}} f_g(D) dD = 1 \quad (5)$$

$$g(t) = \frac{1}{t^*} \exp\left(-\frac{t}{t^*}\right) \quad (6)$$

$$t^* = \frac{W}{F}, \quad F = F_p + F_s$$

Also, from Equation (3), with the boundary condition,

$$D = D_s, \quad t = 0$$

t can be written as Equation (7).

$$t = \frac{D - D_s}{k} \quad (7)$$

Using these equations, the size distribution of product granules can be expressed as Equation (8)

$$f_g(D) = \frac{F_s}{Fkt^*} \int_{D_{min}}^{D} f_s(D_s) \left(\frac{D}{D_s}\right)^3 \exp\left(-\frac{D-D_s}{kt^*}\right) dD \quad (8)$$

The numbers of product granules from the FBG (N) are shown as Equation (9).

$$N = F \int_{D_{min}}^{D_{max}} \frac{f_g(D) \, dD}{\frac{\pi}{6} \rho_p D^3} \quad (9)$$

The numbers of granules in the FBG (N_B) are given as Equation (10)

$$N_B = W \int_{D_{s,min}}^{D} \frac{f_g(D) \, dD}{\frac{\pi}{6} \rho_p D^3} \quad (10)$$

At a steady state, from Equation (1),

$$F_p = k' S_t \quad (11)$$

where S_t is the total surface area of granules in the FBG.

$$S_t = \int_{D_{s,min}}^{D_{max}} \frac{\pi D^2 W f_g(D) \, dD}{\frac{\pi}{6} \rho_p D^3} \quad (12)$$

The growth rate of granules' diameter is shown in Equation (13)

$$\frac{dD}{dt} = k = \frac{F_p}{3W \int_{D_{s,min}}^{D_{max}} \frac{f(D)}{D} dD} \quad (13)$$

From Equation (8) and (13), the size distribution of product granules can be calculated from the size distribution and amounts of the seeds and the mean residence distribution of powder, assuming that layering granulation mechanism is prevailing. Here the assumption is quite acceptable, when $S^* = 0.7$ to 0.8.

Fluidization characteristics

The "granule to granule" agglomeration can occur when S^* is high, e.g., $S^* = 0.9$ to 1.0, which will destroy the smooth tumbling granulation mechanism. The value of S^* is convenient to define the granulation mechanism but the distribution of S becomes sometimes very critical. When $S^* = 0.7$ to 0.8, if there is an extreme S distribution in the granulator, still S^*_{local}, mean value of S in a specific small local volume element, can reach up to 0.9 to 1.0 locally. Therefore, the granule motion in the FBG is very important to reduce the local heterogeneity of S^*_{local}. A force ratio (Φ) in FBG is important, as defined in Equation (14):

$$\Phi = \frac{\text{Forces between individual granules}}{\text{Forces prevailing in FBG}} \quad (14)$$

The motion of granules within FBG can be determined by changing u_0/u_{mf} or adjusting the velocity of gas jet velocity in case of spouted FBG [3].

EXPERIMENTAL

The schematic diagram of an FBG is shown in Figure 3. The configuration of the experimental apparatus (FBG's I.D. 300 mm, 2000 mm in height) is illustrated in Figure 4. The automatic control system of keeping the constant moisture content of granules and maintaining the hold-up amount of granules in the FBG are shown in Figures 5 and 6. The range of the operating parameters are summarized in Table I. The ceramic raw mix, shown in Table II, is employed as the standard sample powder and water used as representative aqueous liquid binder.

RESULTS AND DISCUSSIONS

Influences of Fluidization on Granulation Processes

The "granule-to-granule" agglomeration due to higher S can easily be avoided. However, the small amount of "granule-to-granule" granulation can not always be completely eliminated due to the local S^*_{local} value in the FBG. This amount is called as ξ. The ξ can be controlled by t^* and u_0/u_{mf}. The effect of the hold-up weight of granules (26-62 kg) on the amount of "granule to granule" granulation or ξ is shown in Figure 7. Here the experiment was accomplished at a very high powder feed rate to enhance the effect of granule hold-up on ξ. These results were obtained under the following condition, i.e., the powder feed rate (220 kg/hr), the seed feed rate (10.2 kg/hr), the moisture content of granules (10.2%) and the superficial gas velocity (1.45 m/s). The mean product granule size (D) is approximately 1.4 mm. The formation of ξ decreases with the larger hold-up weight. The effect of the superficial gas velocity (1.14-2.09 m/s) on ξ is shown in Figure 8. Other operating conditions are as follows: powder feed rate: 100 kg/hr; seed feed rate: 12 kg/hr; moisture content: 10.2%; hold-up weight of granule: 44 kg. The relation between minimum fluidization velocities (U_{mf}) and mean granule diameter (D) is shown in Figure 9. These results indicate that the ξ formation does not substantially take place, in case U_0/U_{mf} becomes larger than 2.0. It means that the gas velocity (u_0) should be large enough to avoid the bad motion of granules in fluidization. The mean residence time ($t^* = W/F$) of the feed powder in the FBG is also important. The effect of t on the ξ value is illustrated in Figure 10. The mean residence time must carefully be chosen to avoid the "granule to granule" granulation. This means that the dispersion of liquid binder got worse when the t^* became too short. Within the range of those experiment, t larger than 0.33 hour seems to be necessary. As is seen from Figure 7, the ξ can be affected by changing U_0/U_{mf} and also by selecting the gas volume flow ratio of the gas jet in spouted fluidized beds. In this experiment series, the volume ratio of the gas jet (the volume flow rate of gas jet/total volume flow rate) was chosen not as a variable but as a constant value of 40%.

Size Control of Product Granules by the Seed Control Method

As discussed in the preceding section, the size control of product granules can be attained by the seed nuclei procedure, while the other operating parameters are kept at optimum conditions, maintaining the layering granulation mechanism. However, the determination of the minimum necessary seed diameter ($D_{s,min}$) is indispensable, because the feed seeds can not always be utilized as seed's nuclei but be partially consumed by collision granulation. By using a specific recycle seed and changing the seed mix ratio, the effective seed's size was determined experimentally in the following example.

The operating condition and the number balance of the seed and product granules are shown in Table III. From the data, the minimum seed diameter ($D_{s,min}$) is approximately 0.35 mm in this case. The seeds smaller than 0.35 mm can not effective as be the seed nuclei for the layering granulation but be consumed by the collision granulation. The size of product granule (mean size 1.27-1.68 mm) can be adjusted by the changing the seed's weight percentage (8.5-18%). The size distribution of product granule can be calculated numerically by combining equations (8) and (13), knowing the operation parameters such as the size distribution of seeds ($f_s(D_s)$), feed rates of powder and seeds (F_p, F_s), and hold-up wight of granules in the bed (W), so far as the layering granulation mechanism is predominantly prevailing. The calculated and experimental data are shown in Figure 11, which show good agreement. The minimum necessary seed diameter may be different depending upon each specific granulation process condition, but the basic principle should be identical for any feed materials and type of granulators. The success of continuous 72 hours automatic operation in our pilot plant indicated the verification of the proposed granulation mechanism.

CONCLUSION

A basic idea of controlling the continuous mode granule production system was proposed and experimentally verified by 72 hors continuous operation of fully automated pilot plant consisting of fluidized bed granulator (FBG), seed screens, seed crushers, dryers, recycle seed generating unit, and related necessary sensors and controllers.

In stead of classical, oversimplified batch operation model [10,11], we found an appropriate granulation condition, under which the functions of granule-growth and granule's seed generation can independently be controlled. Although this study's granulator was limited to fluidized bed granulator, the results can fundamentally be applied to almost any other granulators with modest adjustment.

ACKNOWLEDGMENT

This research was supported by Rheopowder Corp.

NOTATIONS

C_1	constant defined in Equation (1)
D	Diameter of granules, [mm]
D_s	Diameter of seed nuclei, [mm]
F	Feed rate of powder including seeds (dry basis), [kg/hr]
F_p	Feed rate of powder (dry basis), [kg/hr]
F_s	Feed rate of seeds (dry basis), [kg/hr]
$f_g(D)$	Size distribution function of product granules, [-]
$f_s(D_s)$	Size distribution function of seeds, [-]
g(t)	Residence time distribution of granules in FBG, [-]
k	Granule growth rate constant defined by Equation (3), [m/s]
k'	Granule growth rate constant defined by Equation (1), [kg/(m² s)]
N	Numbers of granules coming out from the FBG per unit time, [-]
N_B	Numbers of granule in the FBG, [-]
S	Saturation fraction of liquid binder in granules, [-]

S^*	mean value of S in an FBG, [-]
S^*_{local}	mean value of S in a specific volume element, [-]
S_t	Total surface of granules in the FBG, [m²]
t	Time, [s]
t^*	Mean residence time of granule (or powder) in the FBG, [hr]
u_0	superficial gas velocity, [m/s]
u_{mf}	minimum gas velocity for fluidization, [m/s]
W	Hold-up weight of granules in the FBG, [kg]
α	Weight fraction of grown up granules, [-]
β	Weight fraction of growing granules, [-]
γ	Weight fraction of wetted powder and minute granules, [-]
φ	Ratio of adhesion force to hydrodynamic breaking force defined by Equation (14), [-]
ξ	Weight fraction of "granule-to-granule", whose diameter is defined larger than 2.5 time of product mean diameter, [-]

LITERATURE CITED

1. Kono, H. (O.), *Zement-Kalk-Gips* (German), **12**, 549 (1959)

2. Kono, H. (O.), *Chem. Eng.* (Japan), **28**, 709 (1964)

3. Kono, H. (O.), *Chem. Eng.* (Japan), **34** 1035 (1970)

4. Kono, H. (O.), Chem. Eng. (Japan), **43**, 213 (1979)

5. Kono, H. O., *Proceeding of 3rd International Symp. on Agglomeration*, Nürnberg (1981)

6. Kono, H. O., Preceding of International Symposium on Powder Technology (Kyoto, Japan), 625 (1981)

7. Huang, C. C. and H. O. Kono, *Ind. Eng. Chem.*, **28**, 979 (1989)

8. Huang, C. C. and H. O. Kono, *Powder Tech.*, **55**, 19, (1988)

9. Rumpf, H. *Chem. Ing. Tech.* **3**, 144 (1958)

10. Ouchiyama, N. and T. Tanaka, *Ind. Eng. Chem., Process Des. Dev.*, **14**(3), 266 (1975)

11. Kapur, P. C., K. Sastry and D Fuerstenau, *Ind. Eng., Chem. Process Des. Dev.*, 20, 519 (1980)

APPENDIX DERIVATION OF EQUATION (8)

Number of seeds with diameters of D_s to $D_s + \Delta D_s$

$$\Delta N_s(D_s) = \frac{F_s f_s(D_s) \Delta D_s}{\frac{\pi}{6} \rho_p D_s^3}$$

weight of granules out of FBG with diameters D to D+ΔD

$$\frac{F_s f_s(D_s) dD_s}{\frac{\pi}{6} \rho_p D_s^3} \frac{\pi}{6} \rho_p D^3 \frac{1}{t} \exp\left(-\frac{t}{t^*}\right) \Delta t$$

where

$$\frac{\Delta D}{\Delta t} = k, \quad \Delta t = \frac{\Delta D}{k}$$

therefore

$$F_s f_s(D_s) \left(\frac{D}{D_s}\right)^3 \frac{1}{t^*} \exp\left(-\frac{t}{t^*}\right) \left(\frac{D-D_s}{kt^*}\right) \Delta D_s \frac{\Delta D}{k}$$

$$F f_g(D) = \frac{F_s}{kt^*} \int_{D_{s,min}}^{D} f_s(D_s) \left(\frac{D}{D_s}\right)^3 \exp\left(-\frac{D-D_s}{kt^*}\right) dD_s$$

Thus Equation (8) is derived.

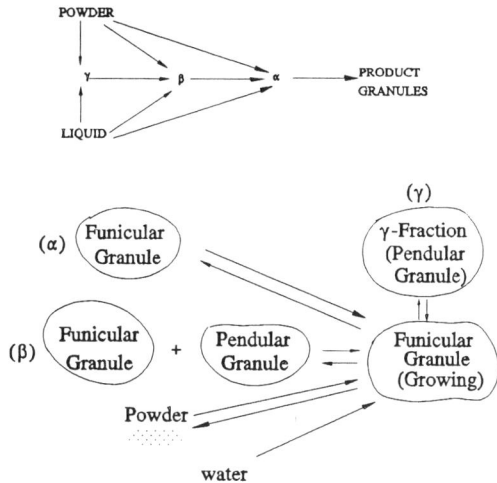

Schematic granulation mechanism α-β-γ model of Kono

Figure 1. Granulation Mechanism in Fluidized Bed Granulators

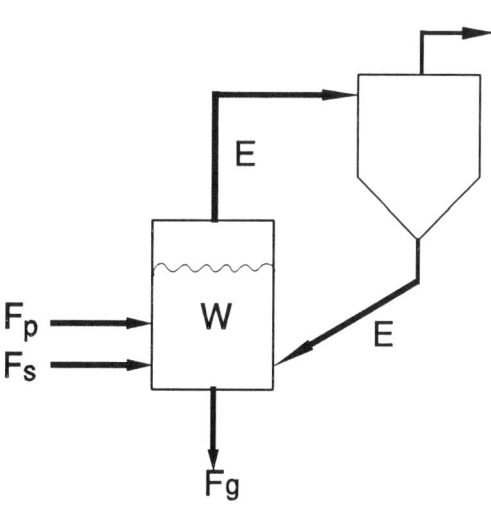

Figure 2. Material Flow Diagram in the Fluidized Bed Granulator

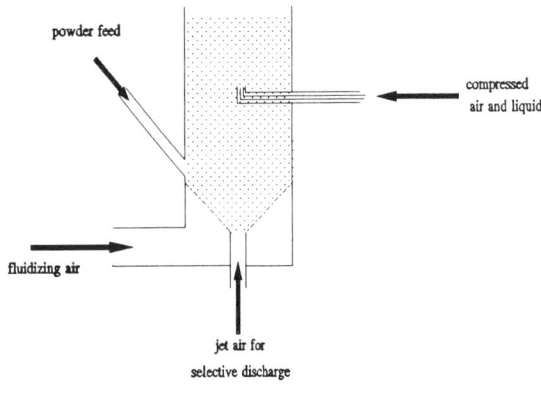

Figure 3. Schematic Diagram of Fluidized Bed Granulators

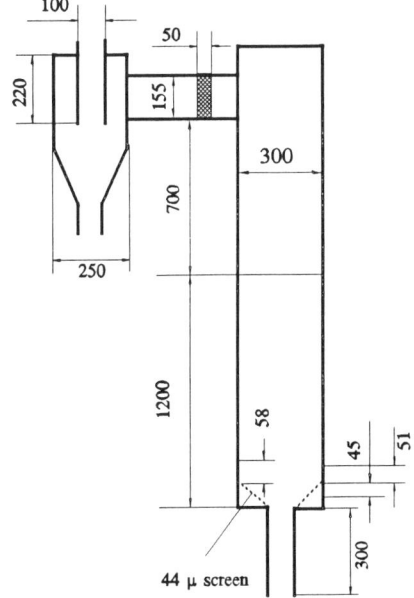

Figure 4. Configuration of Experiment Apparatus

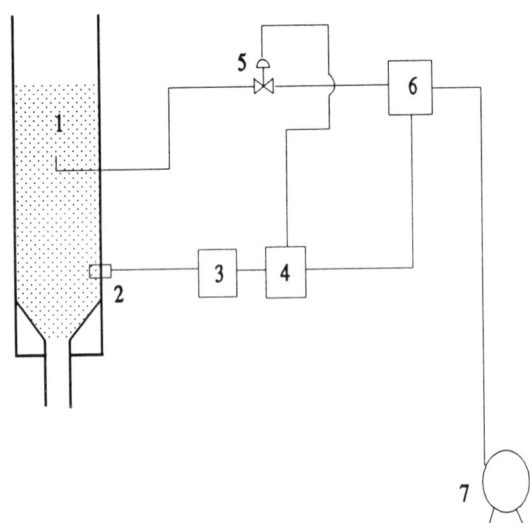

Figure 5. Automatic Control System for the Moisture Content of Granules

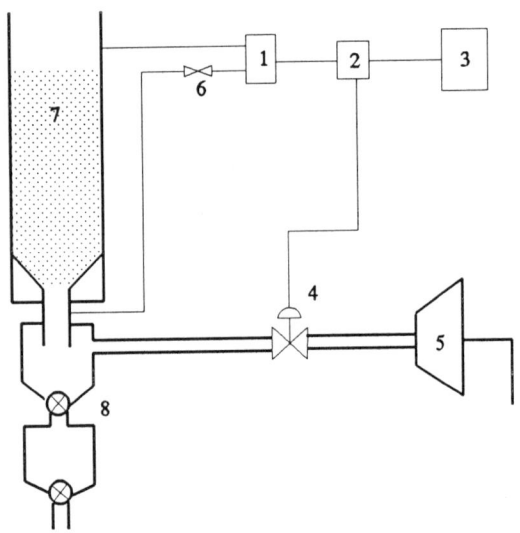

Figure 6. Automatic Control System for the Hold-up of Granules in the Fluidized Bed Granulator

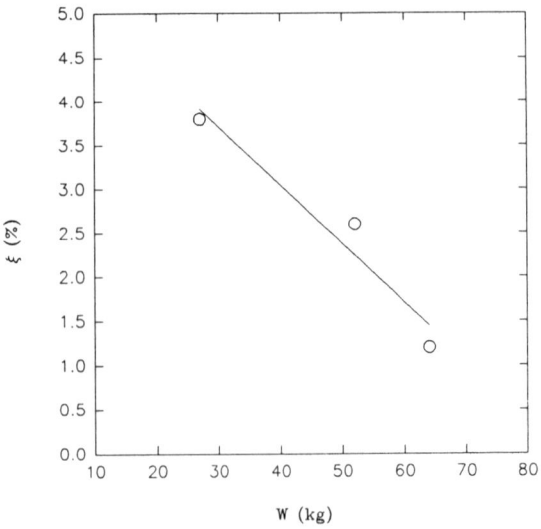

Figure 7. The Effect of the Hold-up Weight of Granules on ξ ($S^*=0.75$, $F_p=220$ kg/hr, $F_s=24$ kg/hr, $u_0/u_{mf}=1.70$)

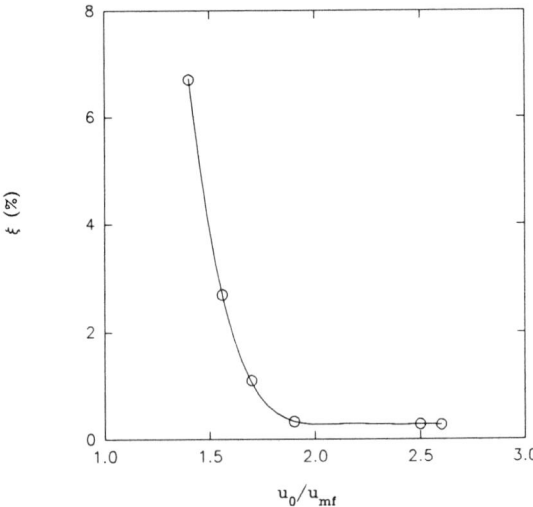

Figure 8. The Effect of U_0/U_{mf} on ξ ($W=44$ kg, $F_p=100$ kg/hr, $F_s=12$ kg/hr, $t^*=0.44$ hr)

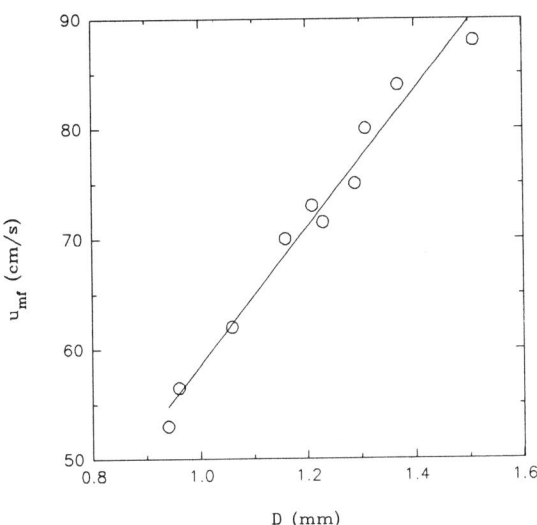

Figure 9. The Relation between U_{mf} and D

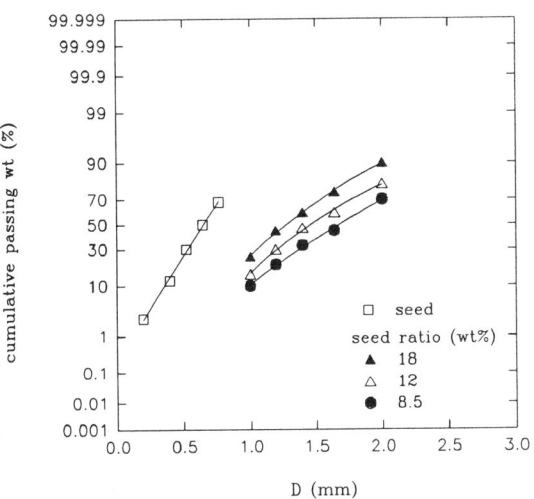

Figure 11. The Comparison of Calculated and Experimental Results on the Size Distribution of Product Granules (detail operation condition is to be referred to Table III)

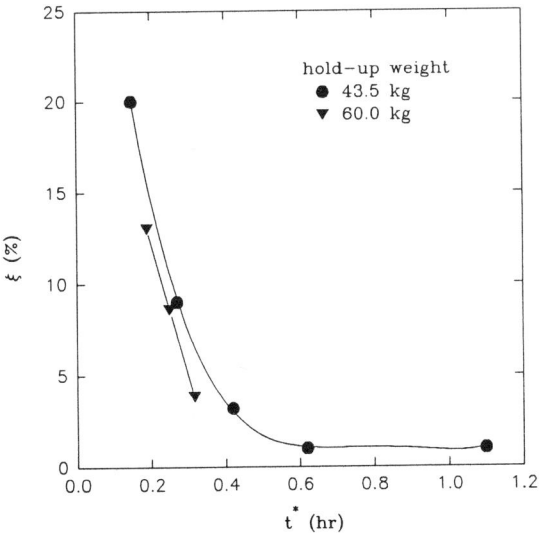

Figure 10. The Effect of t^* on ξ (F_p=100 kg/hr, F_s=12 kg/hr, W=43.5 to 60 kg, u_0/u_{mf}=1.7)

Table I The Range of Designed Operating Parameters in FBG

Parameters	Ranges
Powder feed rate including seeds	40 to 220 kg/hr
Seed feed ratio	5 to 29 wt %
Moisture content	10.2%
Superficial gas velocity	1.14 to 2.09 m/s
Hold-up weight of granules	26.4 to 61.7 kg
Mean residence time of powder in bed	0.26 to 1.07 hr

Table II Properties of Ceramic Raw Mix Powder

Size Distribution		Chemical Composition	
size	weight percentage (wt%)	compound	weight percentage (wt%)
125μ residue	4.2	igloss	26.1
88μ residue	13.8	H_2O	0.8
62μ residue	20.9	SiO_2	28.1
44μ residue	32.8	Al_2O_3	8.0
		Fe_2O_3	2.7
		CaO	31.2
		MgO	1.0
		SO_3	0.3
		total	98.3

Table III The Operating Conditions and the Number Balance of Seeds and Product Granules

The Operating Conditions				
Powder feed rate including seeds			100 kg/hr	
Seeds feed ratio			8.5-18%	
Moisture content of granules			10.2%	
Superficial gas velocity			1.44 m/s	
Hold-up weight of granules in the bed			46.1 kg	
The Number Balance of Seeds and Product Granules				
Seed wt%			8.5	12.0	18.0
cumulative numbers of seeds fed in unit time X 10^3 (1/hr)	size (mm)	1.00	300	220	1030
		0.71	4500	4900	9800
		0.50	10400	14800	25200
		0.35	21700	49600	53500
		0.25	49800	80000	88100
		0.18	190000	337000	371000
number of product granule coming out in unit time X 10^3 (1/hr)			27000	49000	56000

A Two-Stage Fluidized Bed Process for Converting Hydrogen Chloride to Chlorine

M. Mortensen, H.Y. Pan, R.G. Minet and T.T. Tsotsis
Department of Chemical Engineering, University of Southern California, Los Angeles, California 90089
Sidney Benson
Department of Chemistry, University of Southern California, Los Angeles, California 90089
Juan Llibre
Carburos Metalicos, S.A., Barcelona, SPAIN

One of the major waste disposal problems in the chemical industry is the safe and economic handling of large quantities of hydrochloric acid which result from the use of chlorine in many chemical processes. This is a worldwide problem. In the USA, for example, more than 11 million metric tons of chlorine were consumed in 1993 leading to the production of 3 million metric tons of by-product hydrochloric acid. The authors have been working on the development of a new two stage process since 1989 (called the Catalytic Carrier Process) which has reached the pilot plant stage with a two kg/hr plant located at the Zona Franca plant of Carburos Metalicos, S.A. in Barcelona, Spain. The process makes use of fluidized bed reactors and fluidized heat exchangers. As originally conceived hydrogen chloride enters a reactor, where it contacts a fluidized bed of copper oxide supported on a zeolite carrier at 200°C to form a copper chloride complex. From the first reactor the fluidized solid material is carried through unique fluidized bed heat exchangers and strippers and delivered to a second reactor where oxygen is introduced at 360°C to convert the copper complex back to copper oxide, which is then returned to the first reactor. The released chlorine product gas is delivered to a recovery system which includes treatment for separation of residual oxygen, nitrogen, hydrogen chloride and water vapor. Work on this process has been previously reported and the basic technology has been patented in the U.S.A. and Europe. The purpose of this paper is to provide up-dated information for the Catalytic Carrier Process. Recent work at the University of Southern California and Carburos Metalicos have led to significant improvements in the overall process design which will significantly reduce utility requirements by modification of the process steps. These are described in the body of the paper along with economic projections of chlorine recovery costs which are in the range of $80 to $100 per metric ton depending on specific site factors. Laboratory data supporting the new process, called the combination catalytic carrier process is presented.

The process described in this paper is designed to convert waste and by-product hydrogen chloride to chlorine which can then be recycled and reused. When fully developed, this process will enable the economic disposal of hydrogen chloride by regenerating chlorine for return to the chemical process which makes use of it. Among such processes are the production of polyurethanes, isocyanates, titanium dioxide pigments, vinyl chloride monomer, various organic chloride compounds, and many other technologies. A significant reduction in the quantities of chlorine that must be shipped by rail or highway can be achieved by replacing a large portion of the fresh chlorine with recycled product within the operating user plant. Several techniques have been proposed previously for production of chlorine from hydrogen chloride. Such work includes electrolysis, direct oxidation with various inorganic oxidation agents and oxidation of hydrogen chloride by oxygen in the presence of catalyst. These approaches have been reviewed in previous presentations, and publications [1-6]. Among the most interesting is the much studied Deacon process which was patented in 1868 [7]. The basic overall reaction for this process can be written as:

$$2HCl + 1/2\ O_2 \Leftrightarrow Cl_2 + H_2O \quad (1)$$

The catalyst is reported to be copper chloride on a silica or alumina carrier with various promoters including sodium and potassium chloride plus rare earth chlorides. The reaction is exothermic in the normal temperature range of operation, 330 to 380°C, with an equilibrium conversion of hydrogen chloride in the range of 60 to 75% under normal operating conditions. Problems associated with the Deacon approach include vaporization of the catalyst, corrosion in the recovery system and gradual loss in catalytic activity.

The Catalytic Carrier Process described in this paper is based on a detailed analysis of the reaction mechanism for reaction (1) above which suggested a two-step procedure [8]. This mechanism has been reported by Benson and Hisham [9] and can be abbreviated as follows:

$$CuO + HCl \Leftrightarrow Cu(OH)Cl \quad (2)$$

$$2Cu(OH)Cl \Leftrightarrow Cu_2OCl_2 + H_2O \quad (3)$$

$$Cu_2OCl_2 \Leftrightarrow CuO + CuCl_2 \quad (4)$$

$$CuCl_2 \Leftrightarrow CuCl + 1/2\ Cl_2 \quad (5)$$

$$2CuCl + 1/2 O_2 \Leftrightarrow CuCl_2 + CuO \quad (6)$$

Correspondence concerning this paper should be addressed to R.G. Minet.

Consideration of this mechanism leads to the concept of a two-stage process where the reaction between CuO and HCl is carried out in a first stage (the Chlorination reactor) in the temperature range of 100 to 250°C and the oxidation for the release of chlorine is carried out in a second stage (the Oxidizer) at 340 to 380°C. A dual fluidized bed system was selected for this process permitting the chlorine to be carried on the catalyst from the Chlorinator to the Oxidizer, where the chlorine is released and the oxidized form of the solid reactant is returned to the Chlorinator.

THE CATALYTIC CARRIER PROCESS

A conceptual design for the process is shown in Figure 1. The process makes use of two fluidized bed reactors. Hydrogen chloride in gaseous or vaporized state enters the bottom of the first reactor (called the Chlorinator) where it contacts the fluidized particles of active material, which is a zeolite carrying copper chloride, sodium chloride, and various promoters. The metallic content on the carrier is in the range of 8 to 17 weight percent copper. At the reaction temperature of 150 to 200°C the hydrogen chloride reacts with the copper oxide very rapidly in an exothermic step with the temperature being maintained by a tubular heat exchanger generating medium pressure steam. Off gases from the Chlorinator pass through a cyclone dust collector and further treatment. Solids from the Chlorinator carry the copper as oxychloride to a separate external fluidized bed heat exchanger which also serves as a seal with respect to the second reactor, the Oxidizer.

The chlorinated material copper oxide on a carrier flows to the bottom of the Oxidizer where it is contacted with oxygen bearing gas. Although air can be used for this step, the reaction is favored by oxygen concentrations above 50%. The oxidation step is slower than the chlorination step, and in addition, it is endothermic requiring the import of heat through an internally mounted heat exchanger in the fluidized bed. An external fired heater provides the necessary heat input through use of a pumped heat transfer medium. Chlorine is released from the copper impregnated carrier at a temperature of 360 to 380°C. Outlet gas flows through a dust removal cyclone and then passes through cooling and chlorine recovery equipment. The oxygenated solid material overflows from the Oxidizer, through an external heat exchanger which also serves as a seal between the two reactors. It then enters the bottom of the chlorinator to repeat the cycle.

This conceptual design was based on experimental work carried out at the University of Southern California from 1989 until 1993 [5,6]. The experiments were carried out in a 25 mm diameter quartz fluidized bed reactor in a batch cyclic manner as will be described in the next section.

EXPERIMENTAL APPROACH

Catalyst Preparation

The Catalytic Carrier Process involves two distinct gas-solid reactions: the chlorination reaction in which copper oxide supported on an inorganic carrier reacts with hydrogen chloride to produce water and copper chloride and the oxidation reaction in which copper chloride reacts with oxygen to release chlorine and produce an oxide of copper. During the experimental program several different carriers were tested including silica, alumina and zeolite. The copper content was varied from 8 to 17% by weight, and several proportions of promoters were used. In general the preparation followed the procedure given in the next paragraph.

The catalyst was prepared by a wet impregnation method for adding copper chloride (8 to 17% copper by total weight) and sodium chloride (at an equimolar ratio to copper chloride) onto the support material. The resulting slurries were dried at about 100°C until reaching constant weight. The recovered materials thus obtained were screened and then dried at 150°C in a fluidized bed reactor for three hours and then calcined in a N_2 - O_2 atmosphere at 360°C for four hours. After calcining for four hours at 360°C the temperature was raised again to 380°C for several hours. The next step in the catalyst preparation was to chlorinate with hydrogen chloride (typically at 200 to 240°C.) until "breakthrough" of a detectable flow of hydrogen chloride as analyzed by a sodium hydroxide titration. The material was then heated to an oxidizing temperature of 350 to 360°C and an oxidation was carried out for approximately one hour. The sample was then ready for experimentation.

Apparatus

A schematic diagram of the benchtop apparatus is shown in Figure 2. The reactant gases utilized were HCl (technical grade, Matheson), O_2 (ultra-high pure grade, Matheson), and N_2 (ultra-high pure grade, MG). Before being sent to the reactor, the gases from the compressed gas cylinders were treated by passing through purifying columns, which contained indicating drierite ($CaSO_4$) to remove water vapor, molecular sieve and activated charcoal to remove hydrocarbons and any trace of oil and heavy hydrocarbons.

The reactor assembly consisted of a preheater, a quartz tube fluidized-bed reactor and an electric furnace. The preheater was made by using a coil of quartz with an O.D. of 8 mm and I.D. of 6 mm connected to the reactor by fusing it to its base. The reactor was constructed from quartz tubing with an outside diameter of 25 mm and inside diameter of 22 mm with a length of approximately 80 cm. A single-zone tube furnace equipped with Omega-CN-2000 temperature controllers was used to heat the reactor controlled with multi-point chromel-alumel thermocouples inside a thermowell, so that the temperature of the reactor base, center and the top portions were monitored. Uniform temperature profiles along the reactor length were achieved.

EXPERIMENTAL PROCEDURES AND RESULTS

Batch Cyclic Experiments

The first series of experiments were carried out in the following manner: In the chlorination stage, the fluidized bed was heated to the reaction temperature while being purged continuously with a nitrogen stream. When the temperature stabilized, the HCl gas flow was started. The HCl/N_2 gas mixture entered the reactor through the preheater and passed up through the fluidized bed reactor containing the copper oxide bearing reactant. The gas mixture from the top of the fluidized bed reactor then passed through two bubblers in series, which contained a 0.1 M. sodium hydroxide solution to detect hydrogen chloride. The reaction run was continued until hydrogen chloride was detected at the outlet of the reactor, i.e., until breakthrough from the fluidized bed. The oxidation reaction was then started. The chlorine content of the effluent gas was continuously monitored by withdrawing liquid samples from the two bubblers containing a potassium iodide solution to absorb the chlorine. Usually the oxidation step was conducted for 40 to 60 minutes at which time the rate of chlorine evolution slowed to very low levels.

After the oxidation step was completed, the oxygen flow was stopped and the temperature of the reactor was lowered to 150 to 250°C under flow of nitrogen for repetition of the chlorination step. The variables studied were temperature, bed weight, oxygen concentration, oxidation time, the effect of different carriers, and copper metal content on the carrier.

Experiments involving the chlorination step were carried out at atmospheric pressure in a temperature range of 150 to 300°C. For most experiments the total flow rate was 430 ml/min and the inlet hydrogen chloride concentration was 24 Vol. % in nitrogen. The adsorption efficiency was determined by measuring the time for hydrogen chloride breakthrough from the fluidized-bed. The chlorination reaction was very rapid. At temperatures below 250°C, no measurable hydrogen chloride was detected in the reactor effluent prior to breakthrough. Above 250°C, however, increasing the reaction temperature resulted in a significant decrease in the time required to reach breakthrough, and consequently a decrease in the amount of HCl reacted, evidence that the reaction was reversible at higher temperatures. During the chlorination, condensed water vapor was noticed in the exit tubing when the bed was close to breakthrough. These observations are consistent with the reaction mechanism proposed by Benson and Hisham [9].

The experimental results obtained for the chlorination runs suggest that the temperature for hydrogen chloride adsorption should be below 250°C to avoid breakthrough from the reactor bed prior to the complete reaction with the copper oxide present on the carrier.

Since the hydrogen chloride adsorption step is very fast, it was not convenient to measure the reaction rate in the fluidized bed system. Additional studies of the reaction kinetics for this step were conducted in a low pressure reactor system equipped with a UTi 100 C Quadrupole mass spectrometer and are reported elsewhere [5].

Following chlorination, the catalyst was oxidized with enriched air. The minimum temperature at which appreciable oxidation reaction takes place was found to be 280°C. A typical reaction rate vs. time curve obtained for the oxidation step, is shown in Figure 3. The rate is defined in terms of gr of chlorine released per hr per gr of copper on the catalyst per atm of oxygen partial pressure, based on average chlorine evolution per 5 mins time interval and the inlet oxygen concentration. Note that the rate of chlorine evolution falls from approximately 2.5 to 1.25 in the units stated within the first 10 minutes. As the reaction proceeds, the rate of the reaction continues to decrease exponentially until it reaches a value of 0.3 after 30 minutes of reaction. From this point on, the reaction rate continues to fall slowly with time for the specific conditions of the test.

The relationship between oxidation rate and time is related to the decrease in the concentration of active Cu atoms on the carrier surface, available to react as the conversion continues. Cu availability is the controlling factor since the molar flow rate of oxygen is well in excess of the stoichiometric requirements for chlorine evolution. In a dynamic system, where the catalytic carrier is continuously moving thru the reactor, the change in Cu concentration will be significantly less on a time basis. Flow conditions of solid thru the fluid bed would be controlled to maintain the system in the high reaction rate mode. Variations in gas flow rate which changed the fluid-solid particles environment did not result in significant change in the oxidation rate with time as it is shown in Figure 3. This supports the concept that the reaction rate observed is not a strong function of transport conditions.

These results suggest that for the continuous Catalytic Carrier Process solid residence reaction times for the oxidation step of less than 10 minutes are advisable maintaining high Cu availability, to take advantage of the higher initial reaction rate. Overall the oxidation step is slower than the chlorination step and is the rate limiting step for the process.

The effect of temperature on the oxidation rate was investigated in a series of experiments in which the oxidation temperature was varied in the range of 320 to 380°C. Although the observed rate does increase with temperature, this effect becomes less pronounced at the higher temperatures. The experimental results suggest that the optimum temperature for the oxidation step in the process should be around 360°C. In this region the average reaction rate for the first 10 minutes of reaction was 1.8 gr chlorine per gr of copper on the catalyst per hr per atm of oxygen partial pressure. The experiments showed that the oxidation rate increases appreciably with an increase in inlet oxygen concentration up to 60%; thereafter, the effect becomes less pronounced.

Simulation Experiment

The experiments carried out in the total overall cyclic mode provided data which were useful to explore the essential chemistry while giving information that could be used to make preliminary designs for the 2 Kg per hour pilot plant, and for economic evaluation. While the pilot plant was being built some experiments were designed to approach the conditions that could approximate the circulating catalytic two stage system. The procedure for these experiments was as follows:

1. First, chlorinate at a selected temperature, i.e., 150°C, for a specified time, usually 5 minutes using 40 to 50 gr of carrier in the reactor while monitoring the off gas for traces of hydrogen chloride.

2. Then cut-off the hydrogen chloride but continue the flow of diluent nitrogen while rapidly heating the bed to the oxidation temperature, i.e., 360°C, while monitoring the exit gas for hydrogen chloride and chlorine.

3. After reaching the reaction temperature of 360°C (typically in 3 to 4 minutes) the preselected oxygen flow rate was introduced. Oxidation was continued for a total of 10 to 20 minutes, depending on preselected conditions, while the off gas was incrementally analyzed for chlorine with potassium iodide absorption, followed by a sodium thiosulfate titration, and for hydrogen chloride with sodium hydroxide titration every 5 minutes following the chlorine determination.

Steps 1, 2 and 3 were repeated through a series of 10 cycles. A chlorine atom balance was calculated for each cycle. Typical results are given in Table 1 showing the 100 % chlorine atom balances obtained and the indicated conversion of hydrogen chloride to chlorine which averaged between 60 and 80 percent depending on conditions. As stated below the hydrogen chloride conversion can be increased to 100% by an additional chlorination step.

The amount of data collected were too large to be included in this short paper but some generalized conclusions can be drawn:

1. Chlorine recovery as a measure of hydrogen chloride conversion was related to chlorination time with higher recovery resulting from shorter time. The best results obtained for hydrogen chloride conversion were with chlorination temperatures around 100 to 150°C and oxidation times of 15 to 20 minutes at 360°C and 60% oxygen inlet concentration.

2. About 3 to 4 cycles were required to bring the system into 100% chlorine atom balance. In other words, all hydrogen chloride input during the chlorination step reacted 100% with the catalytic carrier as evidenced by the analysis of the off gas. The chlorine which together with the HCl gave a 100% chlorine balance was recovered in the off gas over a 10 to 20 min oxidation time.

3. For this series of experiments, the off gas from the oxidation step contains hydrogen chloride which must be removed in the subsequent chlorine purification step described below.

4. Various strategies were tried which increased chlorine recovery to 80 to 90% but it was concluded that a change in the process sequence was required to achieve essentially an overall 100% conversion of hydrogen chloride, which is necessary to give the most economically viable systems.

Removal of Hydrogen Chloride from the Chlorine Product

Consideration of various alternatives resulted in the design of a series of experiments in which a gas having the composition obtained in the off gas from the oxidation, containing hydrogen chloride, chlorine, oxygen, nitrogen and water vapor would be introduced as the inlet stream to the Chlorinator. Under normal chlorination conditions, at 150 to 200°C, the hydrogen chloride reacted with the catalytic carrier while the inlet chlorine, oxygen, nitrogen and water vapor passed thru to the outlet of the reactor without reacting. After 3 to 4 cycles combined with the oxidation at 360°C, the overall results gave a 100% chlorine atom balance and 99+% hydrogen chloride conversion. These results are shown in Table 2.

From an overall process point of view this modification amounted to using the catalytic carrier chlorination step as a way to recycle hydrogen chloride, appearing in the oxidizer effluent, within the system, giving a product chlorine stream essentially free from hydrogen chloride, although it would contain a significant fraction of water. An analysis of the economic effect of this modification to the Catalytic Carrier Process indicates an increase in the size of the reactors of about 10 to 15 percent compared with the original concept. This would increase capital and operating costs in proportion but the resulting system is still more attractive economically than existing or proposed alternates.

The Combined Process

Another alternative was selected for investigation for experimentation and economic analysis. One major problem with the single stage catalytic oxidation was the limit on hydrogen chloride conversion imposed by equilibrium constraints to approximately 60 to 70% under reasonable operating conditions. By combining the catalytic oxidation reaction as a first stage with the Catalytic Carrier Process as the second stage, it appears to be possible to have a synergistic integration with the off gas from the first stage being purified by the chlorination step in the second stage Catalytic Carrier Process.

A series of experiments were carried out using hydrogen chloride and oxygen as the feed to a fluidized bed reactor operating at 360 to 380°C using the catalytic carrier containing the normal copper chloride resulting from a chlorination step as the catalyst. The results are encouraging as shown in Table 3 but need to be optimized by varying the temperature, ratio of oxygen to hydrogen chloride, bed height and probably using other types of catalyst.

Following these experiments a conceptual gas composition was calculated which combined the output of a Deacon type reactor with the output of the Catalytic Carrier oxidation step to provide a feed stream to a chlorination step. The block flow diagram and flow quantities illustrating these calculations are given in Figure 4. Flow streams for fluidized solids are omitted. Experiments were carried out using the calculated combined overheads as the inlet gas to a chlorinator operating at 200°C with very satisfactory results. All hydrogen chloride was picked up by the chlorination step. These experiments are not conclusive until they are repeated in extensive tests in Process Development Unit with actual circulating catalyst and balanced gas flows. Such tests were started in November 1994 in Barcelona.

OVERALL DESCRIPTION OF THE COMBINED PROCESS

Process Flow Pattern

As shown in the block flow diagram, Figure 4, the total hydrogen chloride entering the combination process would flow thru the first stage combination fluidized reactor along with 90% pure oxygen at a pressure of about 15 psig where a temperature of 380°C would give 70% conversion of the hydrogen chloride to chlorine.

Gas containing mainly chlorine, unconverted hydrogen chloride, residual oxygen, nitrogen and water vapor flows thru heat recovery into the Chlorinator for removal of hydrogen chloride. The hydrogen chloride reacts with the carrier at 200°C which then flows into the Oxidizer after heat exchange. There the fluidized solid is contacted with 90% oxygen to yield, at 360°C, a gas containing chlorine and hydrogen chloride which then flows into the bottom of the Chlorinator along with the overhead gas from the combination reactor. Oxidized catalytic carrier flows from the Oxidizer, after heat exchange. There the fluidized solid is contacted with 90% oxygen to yield, at 360°C, a gas containing chlorine and hydrogen chloride which then flows into the bottom of the Chlorinator along with the overhead gas from the combination reactor. Oxidized catalytic carrier flows from the Oxidizer, thru heat exchangers into the Chlorinator to react with the hydrogen chloride entering that reactor. The chlorine product gas leaving the Chlorinator enters the recovery system to pass thru a dehydration step and chlorine purification for the separation of oxygen and nitrogen depending on the eventual use. By this process, more than 99 % of the original entering hydrogen chloride can be converted and recovered as chlorine.

Energy and Capital Cost Discussion

The two stage combined process will be converting approximately 30% of the total hydrogen chloride in the Catalytic Carrier Process step and 70% in the Deacon step. Since the Deacon type process step is exothermic and operates at a temperature above that in the Oxidizer, where the reaction that takes place is endothermic, the need for a large fired heater will no longer exist saving a significant quantity of fuel and capital investment. Heat available from the chlorination will add to the positive energy flow making the overall operating cost significantly lower than for the Catalytic Carrier Process in its original form.

A very approximate economic evaluation has been made for orientation purposes, which is given in Table 4 for a 60,000 ton per year hydrogen chloride recovery plant with the combination process based on 1994 construction costs. This size plant is consistent with several requests received from potential users of the process.

SCALE UP CONSIDERATIONS

In scaling up this process from laboratory thru pilot plant and eventually commercial operation, there are a number of concerns which must be addressed. The proof of concept unit with 10 inch diameter reactors has been placed in preliminary operation (1994). Results to date show that the reaction rate data obtained are consistent with the information given in this paper, but more experiments are still required. The major difference observed to date seems to be that higher superficial gas velocities are needed to give uniform fluidization in 10 inch reactors as compared with the 1 inch reactors, along with more care in the design of the gas distributing grids.

Other areas requiring investigation are the methods for controlling solids circulation rate, heat transfer characteristics of the fluid beds and the way to optimise the extent of Cu reaction between the oxidation and chlorinating steps.

CONCLUSIONS

This paper has presented experimental data and calculated economic information for the Catalytic Carrier Process and the Combined Process which provides a report on the current status of this development. Overall economics project a chlorine recovery cost in the range of $80 to 100 per ton for a 60,000 ton per year plant which should be attractive to chlorine users having to deal with waste hydrogen chloride disposal.

ACKNOWLEDGEMENT

The financial support provided for this research by Carburos Metalicos, S. A. of Barcelona, Spain and Imperial Chemical Industries of Teeside, United Kingdom is gratefully acknowledged.

LITERATURE CITED

1. Allen, J. A. and Clark, A. J., *Rev. Pure and Appl. Chem.*, **21**, 145 (1971).

2. Arnold, C. W. and Kobe, A., *Chem. Eng. Progr.*, **48**, 293 (1952).

3. Engel, W. F., Waale, M. J., and Muller, S., *Chem. Ind.*, **76** (1962).

4. Feurke, K. H., *Chem. Eng*, **CE41**, March 1968.

5. Pan, H. Y., "Catalytic Carrier Process for Hydrogen Chloride Oxidation", Ph.D. Thesis, University of Southern California, June 1993.

6. Pan, H. Y., Benson, S. W., Minet, R. G. and Tsotsis, T. T., *Ind. Eng. Chem. Res.*, **33**, 2996 (1994).

7. Deacon, H., British Patent 1403 (1868), see also U. S. Patent 85,370, Dec. 1868.

8. Minet, R. G., Benson, S. W. and Tsotsis, T. T., U. S. Patent 4,994,256, Feb. 1991.

9. Benson, S. W. and Hisham, M. W. M., *J. Phys. Chem.*, **89**, 1905 (1985).

TABLE 1

CATALYTIC CARRIER PROCESS
BATCH CYCLIC BENCH-SCALE EXPERIMENTS
AVERAGE OF 30 EXPERIMENTAL RESULTS

	CHLORINATION STEP			OXIDATION STEP	
Flow Rate:	180 ml/min HCl		Flow Rate:	300 ml/min Oxygen	
	300 ml/min Nitrogen			150 ml/min Nitrogen	
Temperature	240°C		Temperature	360°C	
Time	5 minutes		Time	20 minutes	

CHLORINATION	OXIDATION		
Molar Input HCl	Molar Output Chlorine	Molar Output HCl	% Mass Cl Balance
0.042	0.01518	0.01343	100

Experiments in a fluidized bed of 10% copper on zeolite. Total catalyst is 45 g in 25 mm reactor. Each experiment consisted of a chlorination at 240°C for 5 mins, followed immediately by an oxidation at 360°C for 20 mins. Analysis of off gas for chlorine and hydrogen chloride was made every 5 mins.

TABLE 2

CATALYTIC CARRIER PROCESS
REMOVAL OF HYDROGEN CHLORIDE FROM OXYGEN, CHLORINE
AND NITROGEN GAS IN THE CHLORINATOR

Chlorinator Temperature: 200°C
Flow Time: 5 minutes (Repeated 10 times)

		Inlet Flow Rates Milliliters/Minutes		Moles Input	Moles Output
A.	Cl_2	215		0.048	0.048
	HCl	200		0.045	0.001
	O_2	100		0.023	0.023
	N_2	20		0.005	0.005
B.	HCl	200		0.045	0.0004
	O_2	200		0.045	0.045
	N_2	100		0.023	0.023
C.	HCl	200		0.045	0.0004
	N_2	300		0.068	0.068

Experiments in a fluidized bed of 10% copper on zeolite. The total catalyst used is 45 g in a 25 mm reactor. Simulated oxidizer overhead for "A" to show removal of HCl without effect on Cl_2, O_2 or N_2 in gas experiments "B" and "C" show complete removal of HCl for all cases. Each chlorination run was followed with an oxidizer run of 20 minutes at 360°C after which the chlorination was repeated.

TABLE 3

DEACON REACTION DATA
45 GRAMS OF 10% COPPER ON ZEOLITE IN 25 MM DIAMETER FLUIDIZED BED

Flow Rate Ml/Min.	Temp. C	Time Min.	Moles Collected HCl Moles	Chlorine Moles	% HCl Conversion
800 Oxygen 200 HCl	360	15	0.031	0.042	73
600 Oxygen 200 HCl	360	15	0.036	0.041	69
400 Oxygen 200 HCl	360	15	0.044	0.036	62
200 Oxygen 200 HCl	360	15	0.049	0.032	57
400 Oxygen 400 HCl	360	15	0.146	0.038	34
300 Oxygen 180 HCl 100 Nitrogen	360	15	0.042	0.037	64
100 Oxygen 100 HCl 400 Nitrogen	360	60	0.054	0.090	77

The experiment was carried out in a continuous flow manner with samples of gas taken over the time period indicated, after continuous operation for 60 minutes.

TABLE 4

ECONOMIC FACTORS FOR THE COMBINATION CATALYTIC CARRIER PROCESS
CAPACITY 60,000 METRIC TONS PER YEAR OF HYDROGEN CHLORIDE

PLANT INVESTMENT	$MM
Catalytic Carrier Combination, Oxidizer and Chlorinator Unit	5.6
Associated Equipment and Cl_2 Recovery System, Scrubbers, Refrigeration and Storage,	2.8
Waste Heat Recovery, Steam	2.2
Miscellaneous	3.0
TOTAL (Battery Limits Costs)	13.6

OPERATING COST	$/METRIC TON PRODUCT Cl_2
Capital @ 20%	46
Utilities (Fuel, Power and Oxygen)	18
Chemicals and Catalysts	4
Labor and Overhead	6
Maintenance @ 5%	11
	85

FIGURE 1: FLOW DIAGRAM OF CATALYTIC CARRIER PROCESS

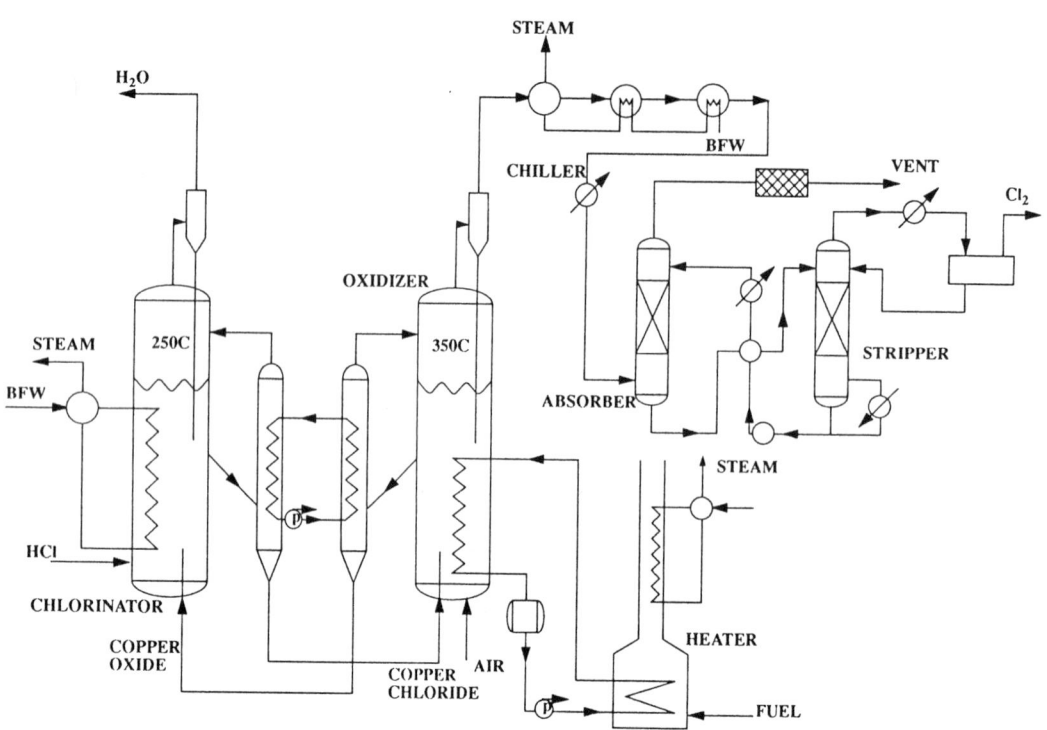

FIGURE 2. SCHEMATIC OF THE EXPERIMENTAL SYSTEM

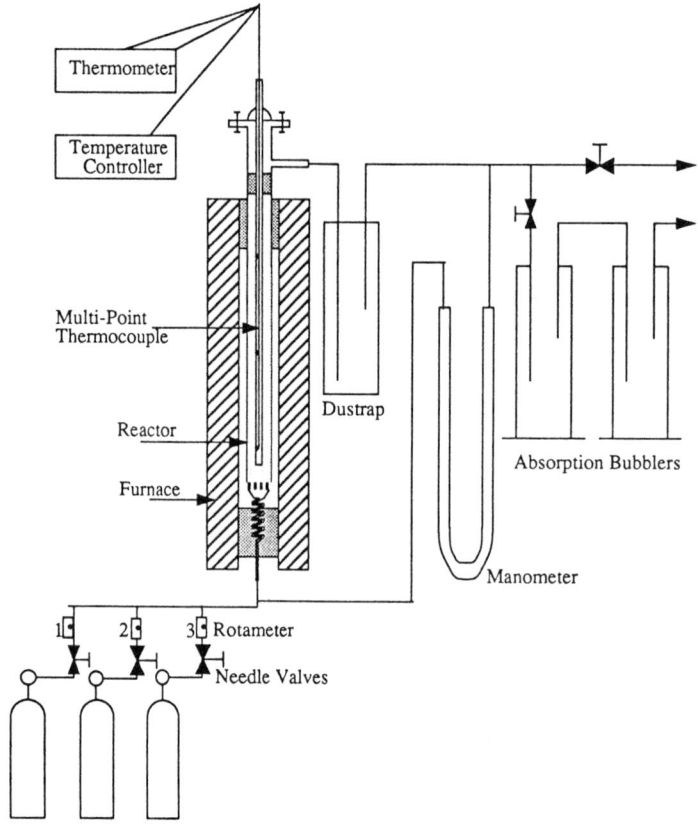

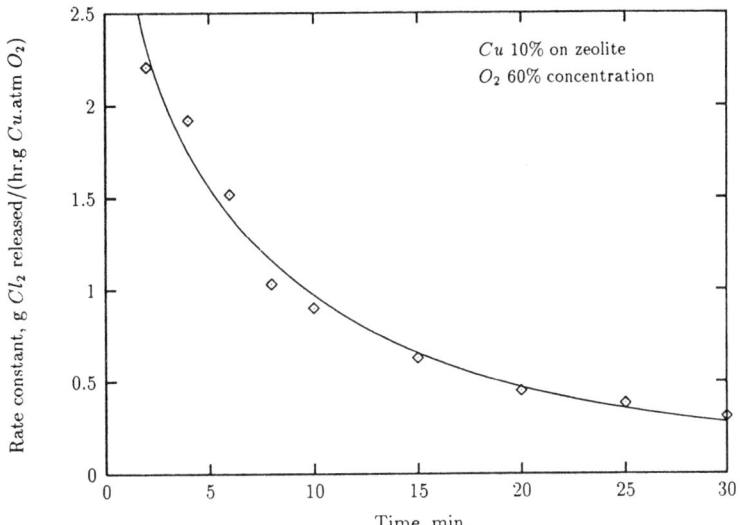

FIGURE 3: REACTION RATE CONSTANT VERSUS OXIDATION TIME. INITIAL CHLORINATION FOR 5 MINUTES AT 240°C AND OXIDATION AT 360°C.

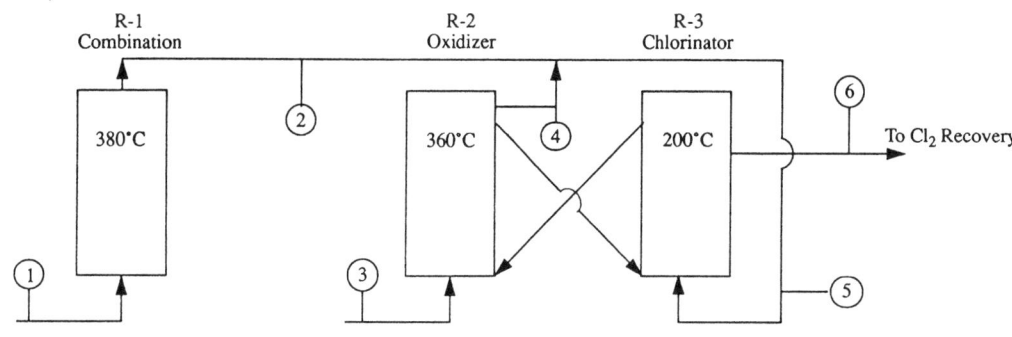

Basis: 1000 kg/hr HCl

STREAM NUMBER		1	1	2	2	3	3	4	4	5	5	6	6
		R-1 INLET	R-1 INLET	R-1 OUTLET	R-1 OUTLET	R-2 INLET	R-2 INLET	R-2 OUTLET	R-2 OUTLET	R-3 INLET	R-3 INLET	R-3 OUTLET	R-3 OUTLET
COMPONENT	MW	KG/HR	KGMOL/HR	KG/HR	KGMOL/HR	KG/HR	KGMOL/HR	KG/HR	KGMOL/HR	KG/HR	KGMOL/HR	KG/HR	KGMOL/HR
HCl	36.46	1000	27.43	300	8.23			99	2.72	399	10.94		0.00
O_2	32.00	308	9.63	154	4.81	131	4.09	66	2.06	120	3.75	220	6.88
N_2	28.02	31	1.11	31	1.11	13	0.46	13	0.46	44	1.57	44	1.57
Cl_2	70.90			681	9.61			292	4.12	973	13.72	973	13.72
H_2O	18.02			173	9.60					173	9.60	247	13.71
TOTAL		1339	38.16	1339	33.36	144	4.56	470	9.36	1709	39.59	1484	35.88

FIGURE 4. MATERIAL BALANCE - COMBINATION PROCESS

The Influence of Hydrodynamics on the Performance of an Interconnected Fluidized Bed System for Regenerative Desulfurization in Coal Conversion Processes

O.C. Snip, R. Korbee, J.C. Schouten and C.M. van den Bleek
Delft University of Technology, Department of Chemical Process Technology, Section Chemical Reactor Engineering, Julianalaan 136, 2628 BL Delft, The Netherlands

An Interconnected Fluidized Bed (IFB) reactor system for regenerative sulfur capture during fluidized bed combustion of coal and coal gasification processes is investigated. A mathematical model is used for an operational analysis of an Interconnected Fluidized Bed (IFB) pilot plant facility. Application of models for the hydrodynamics of the IFB system and sulfur capture and regeneration (based on earlier experimental work) shows that the IFB system is suitable to accommodate a regenerative desulfurization process. The sulfur retention can be effectively controlled by means of the circulation rate of the sorbent that can be best manipulated by variation of the superficial gas velocity in a transport compartment. Also, consequences are given for design and operation of IFB reactor systems.

INTRODUCTION

At Delft University of Technology (DUT) synthetic sorbents were developed for high temperature regenerative removal of SO_2 and H_2S in fluidized bed combustion of coal and coal gasification respectively (Van den Bleek et al. [1] and Wakker [2]). The work presented in this paper is focused on the development of an IFB reactor specifically for regenerative sulfur capture during fluidized bed combustion of coal.

Regenerative desulfurization is advantageous compared to conventional limestone processes with respect to the solids waste stream of the spent sorbent. Another advantage is conversion of the sulfur released from the coal into useful products.

The synthetic sorbent for regenerative SO_2 removal that was developed at DUT consists of 8-9 wt% CaO highly dispersed on a γ-Al_2O_3 carrier. The formation of calcium aluminates enables regeneration at a relatively low temperature of 850 °C. Because this temperature equals the combustor temperature, a highly integrated and energy efficient process is available. The sorbent has been tested extensively for its sulfur capture performance and regeneration activity, stability and mechanical strength in thermogravimetric, fixed and fluidized bed reactors, up to a 1.6 MWth PFBC unit. It appeared that the sorbent is technically suitable for a regenerative desulfurization process and has a lifetime of about 200 cycles of sulfation and regeneration under fluidized bed operation (Van den Bleek et al. [1]).

Next to the additional investments needed for the regeneration step in the process, the make-up of synthetic sorbent is of major importance to the economic viability of the process. The amount of sorbent needed, is determined by deactivation and attrition losses. Sorbent attrition in fluidized beds was shown not to be a major problem. However, the application of conventional pneumatic transport systems (accompanied by high particle velocities of 7-21 m/s) *would* present a major problem, as it results in a high degree of particle breakdown (30-85%). In order to minimize these problems, a new type of reactor is being developed at DUT, based on Interconnected Fluidized Bed (IFB) technology.

In this application the IFB reactor consists of four compartments (Figure 1) and shows a natural compactness due to a lack of particle transport lines. The result is a highly integrated, compact installation which balances the extra costs for regeneration of the sorbent (see Korbee et al. [3] for a preliminary design study of a 100 MWe power plant with regenerative desulfurization, applying IFB technology).

IFB REACTOR SYSTEM

In the IFB reactor, two upflow (lean beds) compartments act as chemical reactors and two intermediate downflow compartments (dense beds) control the transport of solids and may also serve as strippers between the two reactors. Differential aeration is applied to supply the driving force for solids circulation through the system. In turn, the solids flow over a weir and through an orifice which connect the respective compartments. The solids are transported in a fluidized mode with high density and moderate velocity, thus preventing particle break-down caused by high velocity collisions.

In the two upflow compartments (1 and 3), the superficial gas velocity is dictated by the chemical reactions (here: coal combustion/sulfur capture and sorbent regeneration). The second compartment is used to discharge the ash produced in the combustion process, which is separated by means of segregation. The fourth compartment is used to control the Circulation Rate of Solids (CRS) by adjusting the rate of aeration.

Due to the nature of an IFB system the hydrodynamics are of major importance on the reactor performance. This dependency needs careful design and therefore adequate prediction of the performance as a function of the hydrodynamics. In the following sections mathematical models will be developed and applied to predict the CRS and sulfur capture and regeneration performance in the reactor system. With these modelling results an operational analysis (see Figure 3 for the parameters investigated) of the pilot plant facility was performed.

PILOT PLANT FACILITY

To study experimentally the principle of regenerative sulfur capture in an IFB system, a small pilot plant was designed and constructed (Figure 2) at DUT. Coal combustion gas is simulated by feeding a specified mixture of gases at the operating temperature to the reactor. The sorbent is regenerated in a mixture of reductive components (H_2 and CO) in nitrogen. The sizes of the compartments for sulfur capture (1) and regeneration (3) are indicated in Figure 2.

During experiments, the inlet concentrations of SO_2, O_2, H_2O, CO_2, H_2 and CO can be varied and the off-gases of compartments 1 and 3 are analyzed for their composition. Further, the major hydrodynamic parameters can be adjusted: gas velocities, total bedmass (bed height in the compartments), size of the orifice (10-30 mm) and the height of the orifice in the separating wall between the compartments 2 and 3 and between 4 and 1.

MODELLING

In Figure 3 it is indicated how the calculations are performed in modelling the IFB system. Specific sub-models are used to calculate the sulfur retention as a function of hydrodynamic and reaction parameters. In the following sections these sub-models are presented.

1: Hydrodynamics

In the work of Korbee et al. [4] a hydrodynamic IFB model was developed based on force balances in horizontal and vertical direction. This model predicts the solids and gas flow from a dense bed through an orifice into a lean bed. In the dense bed, the axial force balance equals that for the case of solids downflow in a hopper device, except for the aeration. Incorporating a vertical gas pressure gradient resulted in the following relation for the vertical particle stress:

$$\sigma_y = \frac{A_{bd}}{fkP_w}(\rho_p(1-\epsilon_d)g - F(\Delta u_y)) \times (1-\exp(\frac{y-H}{A_{bd}})fkP_w)) \quad [1]$$

In this analysis, it was assumed that part of the weight is carried by the walls of the compartment. In the aerated situation the relative gas velocity (Δu_y) exerts an extra force on the particles and at minimum fluidization conditions the weight of the solids is almost completely carried by the upflowing gas. The fraction of weight sustained by the walls is determined by two parameters: k and f. The relation between vertical (σ_y) and normal stress (σ_N) is determined by k ($\sigma_N=k\sigma_y$) and the friction coefficient f determines the relation between the normal stress and the upward directed stress exerted on the particles ($\sigma_\parallel=f\sigma_N$). The horizontally directed normal stress, at the height of the orifice, is the first part of the driving force for particle flow. The second contribution is the gas pressure drop (ΔP_o) across the orifice. The latter may be a driving or a restricting force depending on its direction. Combining these two contributions to solids flow in a mechanical energy balance yields the solids flow (ϕ_m) through the orifice:

$$\phi_m = C_D A_0 \sqrt{2 \rho_p (1-\epsilon)(k\sigma_y + \Delta P_0)} \qquad [2]$$

Combining Equations (1) and (2) implicitly yields the solids flow through the orifice. To calculate $F(\Delta u_y)$ and ΔP_o, the Ergun equation is applied inserting the relative instead of the absolute gas velocity according to Yoon and Kunii [5]. For a full derivation of the equations above, including experimental verification and model sensitivity analysis, the reader is referred to the paper of Korbee et al. [4].

The model predicts the solids flow (ϕ_m) from the hydrodynamic input parameters: particle properties, IFB size, bed mass and gas velocities. It is important to notice that the hydrodynamic model concerns the solids flow between two adjacent beds. Under the assumption that transport from bed 4 to 1 is controlling (to be verified later), this solids flow equals the CRS through the IFB reactor and can therefore be used in modelling the regenerative desulfurization process.

2: Sulfur capture and regeneration

For the sulfur capture and regeneration processes, it is assumed that the gas and solids phases are ideally mixed. The sulfation is assumed to proceed through the formation of SO_3 according to:

$$SO_2 + \tfrac{1}{2} O_2 \rightleftarrows SO_3$$
$$CaO + SO_3 \rightarrow CaSO_4$$

The sulfation process is further described by an unreacted shrinking core model, the SURE2 model (Wolff et al [6]). In this model, the rate of sulfation is assumed to be first order in the SO_3 concentration at the core radius and first order in the outer surface of the core. The model can adequately predict the sorbent sulfation behaviour for reactors of different scale [6].

The kinetic model for the regeneration is based on two gas-solid reactions and two reversible gas phase reactions:

$$CaSO_4 + H_2 \rightarrow CaO + SO_2 + H_2O$$
$$SO_2 + 2H_2 \rightleftarrows \tfrac{1}{2}S_2 + 2H_2O$$
$$\tfrac{1}{2}S_2 + H_2 \rightleftarrows H_2S$$
$$CaO + H_2S \rightarrow CaS + H_2O$$

The regeneration reactions are considered to be first order in the outer surface area of the sorbent particles and first order in the gas concentrations (like in Schouten and Van den Bleek [7]). The appropriate kinetic parameters were obtained from fluidized bed regeneration experiments [3].

HYDRODYNAMIC MODELLING RESULTS

The model described was used in the operational analysis of the experimental IFB reactor. In Figure 3 the scheme of the calculations is given. First, the CRS is calculated as a function of the hydrodynamic parameters. Then the sulfur capture and regeneration calculations are performed resulting in the overall sulfur retention of the reactor system.

The properties of the particles (SGC-500/sol-gel sorbent) are indicated in Table 1. The sizes of the compartments and orifices in the IFB reactor are indicated in Table 2. The hydrodynamic parameters that were investigated are given in Table 3. The input (hydrodynamic and reaction) parameters are varied in the range of attainable values in the experimental facility. The calculations described were performed with a standard set of input parameters (unless other values are indicated) given in Table 4.

Orifice size.

Increasing the size of the orifice obviously results in higher solids flows. In Figure 4, both the solids flow and orifice gas flow are given as a function of the aeration rate in the dense bed, U_{dense}/U_{mf}.

A maximum solids flow is observed at the conditions of minimum fluidization in the dense bed (note: the bed is at minimum fluidization conditions at $U_d/U_{mf} > 1$ due to gas transfer through the orifice). This is explained by the fact that, with increasing gas velocity, the pressure drop over the bed increases until the minimum fluidization conditions are reached. At higher gas velocity, the porosity increases and gives therefore a lower pressure drop and solids flow. The effect of wall stress ($k\sigma_y$ in

Eq. (2)) is negligible at and above conditions of minimum fluidization. Upon increasing the orifice size, the maximum shifts to higher values of U_{dense}/U_{mf} due to an increasing gas transfer.

The orifice gas flow is seen to be negative (a net gas flow against the solids flow) at low values of U_{dense}/U_{mf}. At a U_{dense}/U_{mf} ratio of 0.75, the amount of interstitial gas carried along with the particles through the orifice (from dense to lean bed) equals the reverse gas transport due to the negative pressure drop across the orifice, resulting in a zero net transport of gas.

Ratio $H_{orifice}/H_{bed}$

The second compartment in the experimental facility is meant for separating and disposing the ash from the sorbent by segregation. For this purpose the orifice can be installed at three different heights. It was calculated that the solids flow decreases proportional to the axial position of the orifice; gas transfer is influenced following the changing pressure drop across the orifice.

Bed area ratio: A_{lean}/A_{dense}

In Figure 5, the bed area ratio (A_{lean}/A_{dense}) is varied for the two combinations occurring in the IFB pilot plant facility as indicated in Table 3. Two effects were observed for the small A_{lean}/A_{dense} ratio: the solids flow starts at a lower U_{dense}/U_{mf} ratio and shows a higher solids flow for the investigated range of U_{dense}/U_{mf}. These effects can be explained by gas transfer phenomena. A smaller A_{lean}/A_{dense} ratio practically implies a smaller lean bed. This will cause the effective gas velocity in the lean bed to increase relatively more resulting in a lower bulk density and therefore a stronger driving force for solids flow.

The assumption that the solids transport between compartments 4 and 1 is rate-limiting is verified with these results. In Figure 5 it can be seen that the assumption holds for equal gas velocities in the dense beds (4 and 2). For equal gas velocities in the connected beds the orifice between compartment 4 and 1 (A_{lean}/A_{dense}=2.3) is rate-limiting for solids circulation as previously assumed. This means that the gas velocity in the second compartment should be set at such value that the solids flow can be controlled in the range needed in the IFB process. In practical situations however (see for example Korbee et al. [3]) this ratio will be relatively large since the dense beds will not serve as chemical reactors. Therefore, the influence of the A_{lean}/A_{dense} ratio will be limited in practice.

Lean bed gas velocity: U_{lean}

A higher lean bed gas velocity results in larger solids flows caused by a higher porosity in the lean bed. This causes a higher pressure drop across the orifice, and subsequently, an increased solids flow. The gas transfer is influenced proportional to the change in pressure drop.

Bed heights: H_{lean} and H_{dense}

The calculations with variation in bed height should be interpreted as a variation in bed mass. Increasing H_{dense} causes a proportional increase in solids flow. For a small H_{lean}, the flow of solids starts at a lower U_{dense}/U_{mf} ratio due to a smaller counter pressure in the lean bed. Increasing H_{lean} shows an equivalent decrease of the solids flow. It is interesting to notice that at a H_{lean} of 0.2 m the dense bed does not become fluidized due to large orifice gas transfer. Geldart and Heasebrouck [8] report similar trends in their experiments. In operating an IFB facility this large gas transfer (H_{lean}=0.2 m) will not occur since the lean beds will be filled at least to the height of the weir (0.3 m).

Orifice shape and type

In their experiments Korbee et al. [4] measured an orifice discharge coefficient of 0.24-0.29. This relatively low value is often found in literature (Martin and Davidson [9], Zhang and Rudolph [10]) and accounts for friction losses in the orifice.

In liquid flow, a C_D value smaller than 1 can be found accounting for friction losses and contraction of the flow from the orifice. Contraction of the solids flow from an orifice in an IFB system is not very likely (see Jones and Davidson [11]). The small value of C_D is usually explained in terms of an empty-annulus effect as observed by Brown and Richards [12]. This effect is a correction for the statistically empty annular zone adjacent to the orifice edge. Further elucidation of the friction that occurs when the solids flow to the orifice and the mechanical energy loss arising from discharge into a bed with fluidized solids compared to open atmosphere is currently investigated at DUT.

Variation of C_D shows a strong (proportional) effect on solids flow but almost no effect on gas flow since this is only determined by pressure drop and orifice size. It can therefore be attractive to use shaped (higher C_D values, Jones and Davidson [11]) orifices if it is desirable to minimize the gas transport relative to the solids flow. This only holds for small scale facilities and/or processes in which the required CRS is not too large. This is obviously caused by the scale effect: for large $D_{orifice}/d_p$ (orifice diameter compared to particle size) the shape of the orifice has a small effect on solids and gas flow (see also Massimila et al. [13])

SULFUR CAPTURE AND REGENERATION

The influence of the major IFB parameters (see Table 4) on sulfur capture and regeneration were investigated. The input parameters are chosen such that regeneration (with hydrogen) of the sorbent is not rate-limiting in the process (i.e., sorbent is completely regenerated). The most important adjustable parameter is the solids flow through the system. For different values of the inlet SO_2 concentration (corresponding to 1-3 wt% sulfur in coal), the sulfur retention (fraction of sulfur captured) was calculated as a function of CRS (see Figure 6). Obviously, sulfur retention increases with increasing sorbent flow. At low CRS the influence of increasing the sorbent flow is strong, while at higher CRS, sulfur retention diminishes to a nearly constant value.

This is explained by the fact the sulfur retention model assumes that the sulfur capture rate depends on the degree of sulfation of the sorbent. Therefore, at low CRS (high sorbent residence times) sulfur retention is relatively low and increases with increasing CRS. At high CRS, the degree of sulfation of the sorbent does not change much with increasing CRS and the sulfur retention is nearly constant.

The influence of the hydrodynamic parameters can be translated to the CRS dependency of sulfur retention, thereby yielding their influence on operation and design of an IFB process for regenerative desulfurization.

CONSEQUENCES FOR DESIGN AND OPERATION OF IFB SYSTEMS

Design

The size of an orifice clearly influences IFB performance: increasing the size of the orifice (increasing CRS while keeping operational parameters constant) will result in a higher sulfur retention, with the previously described limitations at high sorbent conversion. Increasing the height of the inserted orifice will result in a decreasing CRS and consequently a lower sulfur retention. The type and shape of the orifice can have a strong influence on the CRS and therefore on the sulfur retention. Further, the bed area ratio should be considered due to the influence on relative gas transfer and therefore on solids circulation. All these parameters should be carefully chosen in order to fulfil the requirements of the process.

Operation

The increase in bed height (originating from an increase in bed mass) will influence sulfur retention in two ways: 1) the CRS increases and 2) more sorbent material is available for sulfur retention. Bed mass thus has to be considered in operating an IFB system.

The most important operating parameter is the aeration rate of the dense bed used for control of the solids flow rate. Adjusting the aeration rate, the CRS and therefore the sulfur retention can be controlled. In principle, the CRS is best controlled by operating the dense bed in the defluidized regime. In this regime the solids flow is less dependent on the orifice gas pressure drop and therefore offers a better way to control the solids flow rate.

CONCLUSIONS

A mathematical model was used for an operational analysis of the IFB pilot plant facility at DUT. The influence of hydrodynamic parameters on the CRS and therefore on the sulfur retention has been determined. Further, some consequences are given for design and operation of IFB systems in general.

The size of the orifice greatly influences the CRS and orifice gas transfer. The size should be chosen in such a way that the obtainable range of the CRS is sufficient for the requirements of the process.

A careful look at the modelling results reveals that operation of an IFB system should be done in such a way that the aeration rate to one of the transport compartments is the control parameter in the process. This is possible when the design accounts for the ratio of bed areas, lean bed gas velocity, position and size of the orifice. This can be verified by using the IFB simulation model, as presented, which is therefore a useful tool in designing IFB systems. This will allow optimal design of future facilities of commercial scale.

Modelling the DUT IFB pilot plant facility further indicates that the IFB system may be effectively used in a regenerative desulfurization process. In forthcoming experimental work these results will be verified.

ACKNOWLEDGEMENT

The authors wish to acknowledge the financial support from the Commission of European Communities (contract no.JOUF-CT91-0063) and the Delft University Stimulation fund 'Beek' (contract no. 1-71925).

NOTATION

A	cross sectional area	[m^2]
C_D	discharge coefficient	[-]
d,D	diameter	[m]
f	friction factor	[-]
$F(\Delta u_y)$	local pressure gradient	[N/m^3]
g	gravitation constant	[m/s^2]
H	bed height	[m]
P_w	wall perimeter	[m]
Δp_o	orifice pressure drop	[Pa]
U	superficial gas velocity	[m/s]
u	interstitial (linear) velocity	[m/s]
Δu	relative velocity	[m/s]

subscripts
bd	bed
h	hydraulic
mf	minimum fluidization
N	normal
o	orifice
w	wall
lean	lean bed, operated at relatively high gas velocity
dense	dense bed, operated at relatively low gas velocity

Greek
ε	bed void age	[-]
ρ	density	[kg/m3]
σ	stress	[Pa]

LITERATURE CITED

1. Van den Bleek, C.M., Duisterwinkel, A.E., Frens, G., Gerritsen, AW., Korbee, R., Lin, W., Schouten, J.C., Verheijen, P.J.T., Wolff, E.H.P.,"Regenerative Desulfurization in Fluidized Bed Combustion of Coal" Final Report of C.E.C. Contract no. EN3F0014-NL, Delft University of Technology, The Netherlands (1992).

2. Wakker, J.P., "Development of a high temperature steam regenerative H2S removal process based on alumina supported MnO and FeO", Ph.D. thesis, Delft University of Technology, The Netherlands (1992).

3. Korbee, R., Grievink, J., Schouten, J.C., Van den Bleek, C.M., "Preliminary design of a 100 MWe power plant with regenerative desulfurization, applying Interconnected Fluidized Bed Combustion", Proc. 12th Int. Conf. on Fluidized Bed Combustion, San Diego, USA, 1143 (1993).

4. Korbee, R., Snip, O.C., Schouten, J.C., Van den Bleek, C.M., *Chem. Eng. Sci.* **49** (24B), 5819 (1994).

5. Yoon, S.M., Kunii, D., *Ind. Eng. Chem. Proc. Des. Dev*, **9**, 559 (1970).

6. Wolff, E.H.P., Gerritsen, A.W.G., Bleek van den, C.M., *Can. J. Chem. Eng*, **71**, 83 (1991).

7. Schouten, J.C., Van den Bleek, C.M., "The DUT SURE model: A simple approach in FBC Sulfur Retention Modelling", Proc. 9th Int. Conf. on Fluidized Bed Combustion, Boston, USA, 749-761(1987).

8. Geldart, D., Haesebrouck, M., *Chem. Eng. Res. Des.*, **61**, 224 (1983).

9. Martin, P.D., Davidson, J.F., *Chem. Eng. Res. Des.*, **61**,162 (1983).

10. Zhang, J.Y., Rudolph, V., *Ind. Eng. Chem. Res.*, **30** (8), 1977 (1991).

11. Jones, D.R.M., Davidson, J.F., *Rheol. Acta*, **4** (3), 89 (1965).

12. Brown, R.L., Richards, J.D., *Trans. Inst. Chem. Eng.*, **37**, 243 (1960).

13. Massimila,L, Betta,V, Rocca Della, C, *AIChE J.*, **7** (3), 502-508 (1961).

Table 1. Particle (SGC 500 sorbent) and gas properties (at 850 °C).

Parameter/property	Value and units	
ρ_p	1400	kg/m³
d_p	2	mm
ϵ_{fix}	0.45 (fixed bed porosity at 20°C)	
ρ_{bulk}	770	kg/m³
ψ	0.98 (particle sphericity)	
U_{mf}	0.8 (at 20°C)	m/s
	0.65 (at 850 °C)	m/s
α_r	24 (angle of repose)	°
Ca wt %	8.91	
ρ_{gas}	0.317	kg/m³
η_{gas}	4.75E-5	Pa s

Table 2. Sizes in IFB reactor/pilot plant facility.

Compartment size (m²)	D_h (1) (m)	Weir length (m)	$D_{orifice}$ (mm)	A_{lean}/A_{dense} (-)	$H_{orifice}$ (2) (mm)
1. 0.14x0.14	0.14	1→2=0.14	10	4→1=2.33	41
2. 0.06x0.14	0.084		15		111
3. 0.06x0.06	0.06	3→4=0.06	20	2→3=0.43	181
4. 0.14x0.06	0.084		25		
			30		

(1) D_h = 4 x (surface of the compartment)/(perimeter of compartment)
(2) $H_{orifice}$ = measured from distributor plate and
H_{weir} = 0.3 m

Table 3.
Hydrodynamic parameters used in modelling in the indicated range.

Parameter		Range/value	
$D_{orifice}$	size of the orifice	10 - 30	(mm)
$H_{orifice}$	height of the orifice	41, 111, 181	(mm)
H_{bed}	bed height of dense bed	0.27 - 0.33	(m)
	bed height of lean bed	0.2 - 0.3	(m)
U_{dense}	gas velocity in dense bed	0 - 2	(U_{mf})
U_{lean}	gas velocity in lean bed	1.5 - 2.5	(U_{mf})
A_{lean}/A_{dense}	bed area ratio	0.5, 2.3	(-)
C_D	discharge coefficient	0.2 - 0.5	(-)

Table 4.
Standard set of operational properties used in calculations

Parameter		Standard value	
$f_s = f_d$	static and dynamic friction factor	1	(-)
C_D	discharge coefficient	0.5	(-)
$H_{dense} = H_{lean}$	dense and lean bed height	0.3	(m)
$H_{orifice}$	height of the orifice	0.14	(m)
$D_{orifice}$	diameter of the orifice	20	(mm)
A_{lean}/A_{dense}	bed area ratio	2.3 (4→1)	(-)
U_{lean}	lean bed gas velocity	2	(U_{mf})
C_{SO2}	SO_2 inlet concentration	0.1	(v%)
C_{O2}	O_2 inlet concentration	4	(v%)
C_{H2}	H_2 inlet concentration in regenerator	10	(v%)

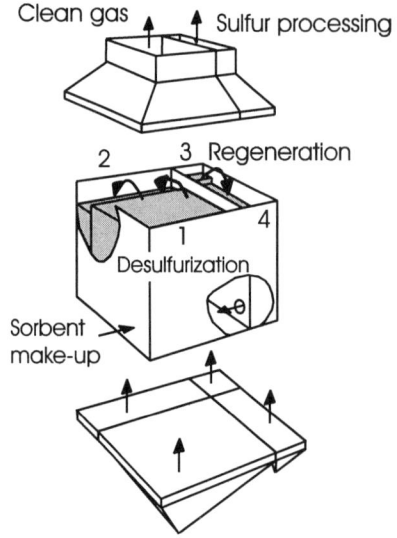

Figure 1. *IFB reactor set-up for regenerative desulfurization.*

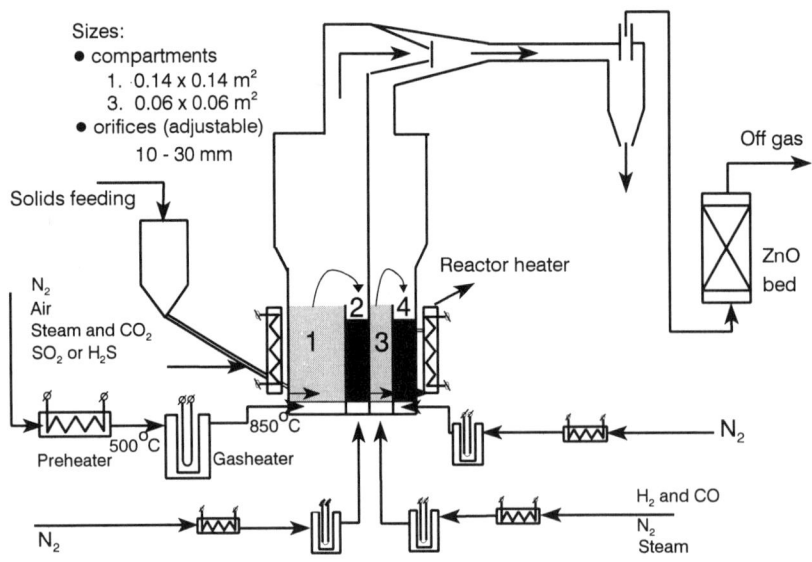

Figure 2. *Schematic view of the IFB pilot plant.*

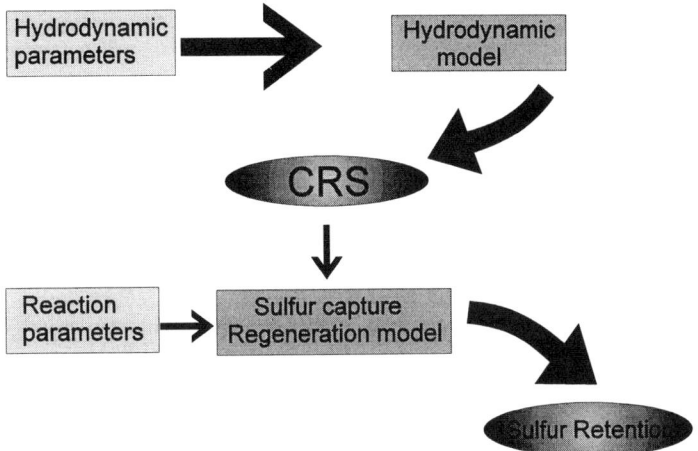

Figure 3. Calculation scheme for the IFB system.

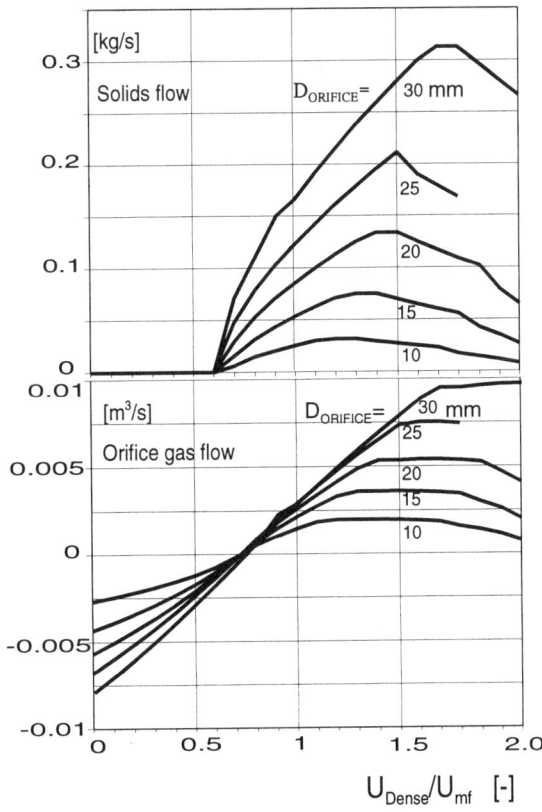

Figure 4. Solids and gas flow vs. gas velocity in dense bed, for indicated orifice diameters.

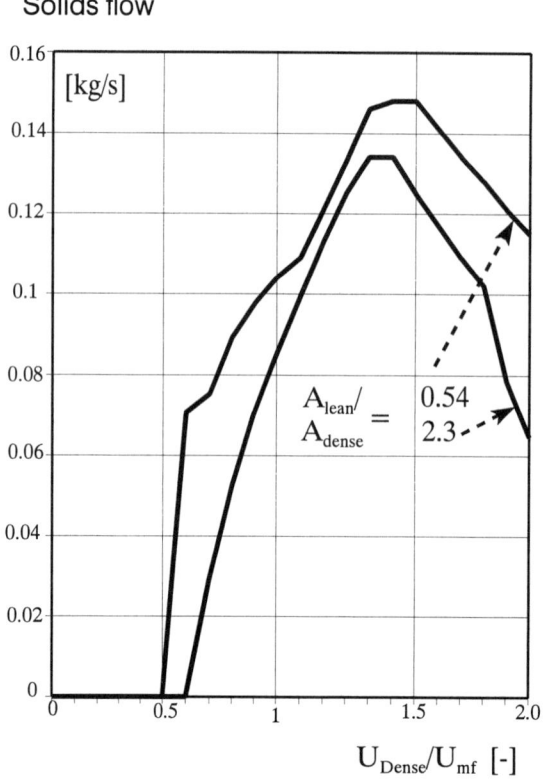

Figure 5. Solids flow vs. gas velocity in dense bed for indicated A_{lean}/A_{dense} ratio.

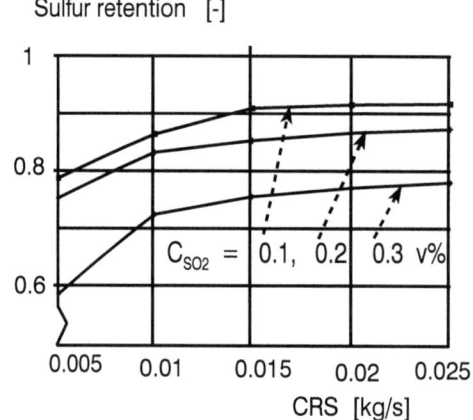

Figure 6. Calculated sulfur retention versus CRS, in IFB pilot plant.

Measurement of Bottom Bed and Transport Disengagement Heights in Beds of Fresh FCC Catalyst

D. Geldart, Yan Xue, and H.-Y. Xie
University of Bradford, Bradford, West Yorkshire, BD7 1DP, UK

The fluidization of FCC powder has been studied at velocities up to 1.2 m/s using instantaneous pressure signals in a fluidized bed 290 mm I.D. and 5.3 m overall height with an internal cyclone. The pressure measurements were used to determine the height of three zones-dense bed, transport disengagement zone, and dilute phase. Measured values of the transport disengagement height are compared with predictive correlations in the literature. Correlations which require a knowledge of bubble size and/or velocity at the bed surface best describe the experimental results. The heights of the three zones changed gradually with gas velocity with no indication of a change in flow regime.

Most fluid bed catalytic reactors making products such as acrylonitrile, phthalic anhydride, and chlorinated hydrocarbons, as well as FCC regenerators, operate at superficial velocities around 0.6 to 0.8 m/s. Because of the size and density of the powder, (generally in the range 50-70 μm and 1300-1800 kg/m^3 respectively), the entrainment rate is very high and the entire bed circulates through the cyclones many times per hour. The circulation rate depends not only on gas velocity, but also on the distance between the top of the dense bed and the entrance to the cyclones- the *freeboard*. Ideally, to minimise entrainment, this distance should be larger than the transport disengagement height, and one of the objectives of this work was to measure bed expansion and TDH in a reasonably large bed using a typical fluid bed catalyst. Several of the existing correlations for TDH rely on a knowledge of the size and/or velocity of bubbles at the surface of the dense bed and there is now firm evidence for the existence of an equilibrium bubble size in Group A powders. Independent measurements of the equilibrium bubble sizes and velocities have been made recently in our laboratories by Xie [1] thereby offering an opportunity to compare correlations realistically with reliable experimental data.

Although it is widely believed that there is a critical velocity at which bubbling breaks down into so-called *turbulent fluidization*, the evidence is conflicting (e.g. Rhodes and Geldart, [2]; Grace and Brereton [3]). Newton [4] recently showed X-ray movie film of a fine silica gel (a Group A powder) fluidized at increasing total pressure, and this certainly revealed a gradual reduction in bubble size, to the point where, at very high pressure, bubbles disappeared altogether. If there is a transition from bubbling to a different sort of (turbulent) fluidization at ambient conditions then a significant change in bed density, entrainment rate, or TDH should be observable as the gas velocity is increased across the range at which transitions are said to occur, and within which most commercial catalytic units operate.

The arrangement of our rig, in which all entrained solids are returned to the dense bed, allowed us to make relevant observations of zone height without the confusion which is present when there is an expanded upper section which allows accumulation of powder within it at the expense of the dense region.

EXPERIMENTAL EQUIPMENT

Experiments were carried out using a fluidized bed of 290 mm internal diameter and 5.3m overall height. The gas distributor was a plate drilled with 9mm diameter holes on a triangular pitch of 16.6mm. To prevent particles from falling through the holes, a fine mesh (250 wire/inch) with a filter paper covered the plate. The air supply was provided by a Roots-type blower. The flow rate was controlled manually using a blow-off valve and was measured using an orifice plate. In order to simulate the arrangement in commercial catalytic reactors, an internal cyclone was used to collect the entrained solids and return them to the bed.

Thirteen pressure taps were mounted between 25mm and 5025mm above the distributor. Thirteen Sensym differential pressure transducers type 143SC01D (±1psi) were used to measure the pressure drops between the pressure taps, and a type 142SC05D (0-5psi) was connected to the bottom pressure tap to measure the overall pressure drop across the bed. Both types of pressure transducer were temperature-compensated and signal-conditioned and had very low dead volumes. The signals from the transducers were sampled, and collected and stored on a floppy disk under controlled sampling frequency and sampling time, using an Amplicom Live Line PC30 board built into an Elonex PC-386SX computer.

All the pressure taps were purged at a line velocity of 2m/s to prevent blockage by the powder. The experimental apparatus and the arrangement of the pressure taps are shown in Figure 1.

A fresh FCC powder (Quantum 2000 supplied by Crosfields Catalysts) with a mean size of 58µm was used in the experiments. The properties of the powder are given below:

ρ_P (kg/m^3)	$\bar{d}_{sv}$ (µm)	U_{mb} (cm/s)	U_{mf} (cm/s)	V_t(calc.) (cm/s)
1320	58	0.46	0.11	9.3

ZONE DEFINITION

Li and Kwauk [5] have shown that in a high velocity circulating fluidized bed the cross-sectional mean voidage exhibits two zones of nearly constant void fraction, separated by a transition zone with an inflection point. Some researchers, e.g., Sciazko et al [6] used the average of visually-observed maximum and minimum bed heights of the fluctuating bed surface to obtain the bottom bed height. Using a coarse sand in a large circulating fluidized bed, Johnsson et al [7] defined zones based on the measured pressure drop profile along the column-the height of the dense zone, H, was defined as the height at which the pressure drop started to deviate from a straight line; and an exponential function was used to fit the data in the transport disengaging zone.

Figure 2 is an example of the measured pressure drop profile along the column at a superficial gas velocity of 0.65m/s. It can be seen that the height of the bottom, or dense, bed can be obtained by the method proposed by Johnsson et al [7]. Although the height of the transport disengaging zone (TDH) could be obtained from the height at which the pressure drop coincides with the nearly-horizontal pressure drop line, in practice this is not suitable since, for better accuracy, it requires more measuring points more closely spaced; when adjacent pressure taps are too close the pressure drops are only a few millimetres water gauge and this may fall within the zero point range of a pressure transducer.

However, the three zones can be easily seen and obtained from a plot of the pressure drop gradient, dP/dH, (essentially the bed density), versus bed height, as shown in Figure 3. The two near-horizontal lines, representing the bed densities of the dense and dilute phases, are clearly separated so that it is easier to obtain the effective heights of the dense phase and the transport disengaging zone by drawing a line through the measured data to meet the two horizontal lines. All the experimental results of the heights of the zones reported below are obtained by this method.

EXPERIMENTAL RESULTS

Dense-phase Height

Experimental values of the dense phase height are plotted in Figure 4 in dimensionless form as H_B/H_{mf} vs. $U-U_{mf}$ for superficial gas velocities up to 1.2m/s. The figure shows that the bottom bed height increased by about 5% above H_{mf} for superficial gas velocities up to 0.4m/s, and then gradually decreased to about 70% of H_{mf} at 1.2m/s, which is about 13 times the terminal velocity of the mean particle size.

Transport Disengagement Height (TDH)

The transport disengaging height (TDH) and the change in suspension density in this zone are useful parameters for design and operational purposes since they influence the rate of entrainment (and hence the rate of internal circulation of both catalyst and product gas (Geldart et al. [8])) and the heat transfer coefficient between bed and heat transfer surfaces near and above the dense zone. Consequently the TDH has been the subject of a number of studies which have resulted in several empirical correlations.

The most common method of measuring TDH has been to measure the entrainment rate in batch or semi-continuous experiments at various velocities, and then to change the position of the gas offtake (e.g. Chan and Knowlton. [9], Schuurmans [10]). Visual observation has also been used, and Sciazko et al [6] measured TDH visually for coarse particles as the height above which no downward moving particles were observed. This technique cannot be used for FCC or similar fine powders because of the obscuration which occurs at the wall.

Fournol et al [11] developed an on-line method to measure TDH using a particle sampling technique. Instead of changing column height, powder sampling was done at a number of heights along the column to measure the solids entrainment rate and particle size distribution up the column.

In our work, TDH is obtained from the curve of the pressure drop gradient (i.e. suspension density) versus bed height, based on the measured pressure drop profile, as described earlier.

Experimental results are plotted on Figure 5 for gas velocities up to 1.2m/s, and are compared with the Zenz correlation [12] in Figure 6. The measured TDH values presented in this format fall between the curves for bubble sizes between 2.5 and 7.6cm at superficial gas velocities up to 1.2m/s. This is in agreement with an equilibrium bubble size of 4.8cm predicted by the Geldart equation [13]:

$$d_b = \frac{2}{g}(V_t')^2 \quad (1)$$

where g is the gravitational acceleration, and V_t' is the terminal velocity of particles which have a size 2.7 times the mean size of the powder. This also agrees well with values measured by Geldart and Xie [14] and Xie [1] using pressure probes, which showed a maximum bubble size of about 4cm for their FCC powders which had mean sizes in the range 26 to 60μm. Hatano and Ishida [15] measured a bubble size of 6cm for their coarser (75μm) FCC at velocities up to 0.3 m/s, so these values are all consistent with each other.

CORRELATIONS FOR TDH IN THE LITERATURE

Wen and Chen [16] proposed that the TDH can be obtained from the entrainment rate at the bed surface E_0, and the elutriation rate, E_∞, by assuming that the TDH is reached when the entrainment rate has fallen to 1% above the elutriation rate. They give:

$$TDH = \frac{1}{a}\ln\left[\frac{E_0 - E_\infty}{0.01 E_\infty}\right] \quad (2)$$

where a is a constant and a value of 4m^{-1} is recommended. The entrainment rate at the bed surface is given by:

$$\frac{E_o}{A \cdot dBo} = 3.07 * 10^{-9} \frac{\rho \cdot g}{\mu^{2.5}} (U - U_{mf}) \quad (3)$$

where A is the cross sectional area of the column, d_{B0} is the bubble size at the bed surface, ρ is gas density, and μ the gas viscosity. The elutriation rate, E_∞, can be calculated from Equation (4):

$$\frac{E_\infty}{\rho U} = 23.7 \exp(-5.4 \frac{V_t}{U}) \quad (4)$$

For conditions in which the particles are so fine that they are all entrainable at the operating velocity, V_t is the terminal velocity corresponding to the mean size of the powder.

Horio et al [17] proposed a more simple expression for the TDH based on the bubble size at the bed surface:

$$TDH = 4.47 d_{B0}^{0.5} \quad (5)$$

However, Baron et al [18] claimed that the TDH should be a function of the absolute bubble velocity rather than bubble size and proposed:

$$TDH = 0.22 U_{bs}^2 \quad (6)$$

where U_{bs} is the absolute bubble rise velocity. According to Xie [1], the bubble velocity in beds of FCC is given by:

$$U_{bs} = U - U_{mf} + 1.5\sqrt{g d_b} \quad (7)$$

Fournol et al [11] found, using FCC powder, that the solids entrainment rate and particle size become constant when:

$$TDH = \frac{1000 U^2}{g} \quad (8)$$

Another empirical expression for TDH was proposed by Amitin [19]

$$TDH = 0.85U^{1.2}(7.33 - 1.2\log U) \quad (9)$$

The approach of Pemberton and Davidson [20] is based on the decay of turbulence in the freeboard caused by so-called 'ghost bubbles' from the fluidized beds. In common with Wen and Chen, they define TDH as the height above the bed surface at which 99% of the particles have disengaged, and this leads to the expression:

$$TDH = -(1/\beta)\ln[1 + (1/K)\ln(0.01 + 0.99\exp(-K))] \quad (10)$$

where β is a turbulence decay constant correlated by:

$$\beta = 230\sqrt{\mu/\rho}.dBo^{-0.45} \quad (11)$$

and:

$$K = \frac{0.71 dBo.u_o'}{\beta D^2 (U - Vt)(1 + S/12)} \quad (12)$$

S is the Stokes number $\rho_p d^2 U/18\mu D$, and u_o' is the fluctuation in the freeboard gas velocity just above the bed surface, which must be related to the bubble velocity. Pemberton and Davidson provide no correlation for u_o', but they make the reasonable assumption that the fluctuations are equal to the bubble velocity at the bed surface, U_{bs}.

DISCUSSION

The comparison between the experimental results and these empirical correlations is shown in Figure 7. The figure shows that both the Wen and Chen and the Baron correlations give satisfactory predictions when used in conjunction with the bubble size (4 cm) and bubble velocity Equation (7). The agreement is also quite good when using the equilibrium bubble size predicted from Equation (1). The Horio correlation gives a constant TDH of about 1m at all superficial gas velocities for a constant bubble size. It also turns out that, for constant bubble size, the Pemberton and Davidson equation is virtually independent of gas velocity since the β term dominates in Equation (10). Both the Amitin and the Fournol *et al.* correlations give predictions which are much too high. The correlation, which best fits our results is the empirical graphical one proposed by Zenz [12].

However, there is an important question as to whether these results, obtained in a reasonably large diameter column (0.29m), or the existing correlations, allow us to predict TDH in columns of industrial size operating at high pressure and/or temperature. Chan and Knowlton [9] working with Ottawa sand approx. 100μm in mean size, found that the Zenz correlation fitted their data best for ambient conditions, but that TDH increased significantly with increasing pressure; only the correlation of Frantz and Juhl [21] predicts this trend. Few experimental data from large units have been published, but those that have (Wolfarth [22] and Schuurmans [10]) show that both entrainment rates and TDH increase as vessel diameter increases. The earliest version of the Zenz plot, (Zenz and Weil [23]) in which column, not bubble diameter is a parameter, recognises this. Correlations for TDH based on bubble rise velocity can potentially predict that TDH increases with column diameter if the bubble velocity is calculated using Werther's [24] formula:

$$U_{bs} = \varphi\sqrt{g.db} \quad (13)$$

where, for Group A powders, $\phi = 2.5D^{0.4}$ for 0.1<D<1m, and $\phi=2.5$ for D>1m diameter.

Another factor, not studied here but referred to in Schuurman's paper and in the data of Geldart and Pope [25], is that TDH is a function of particle size. The concentrations of the finest fractions reach an equilibrium value quite rapidly (small TDH) whilst those of the coarser, but still entrainable fractions continue to decline up to the column exit.

Finally, it is worth pointing out that in no published results is there any evidence of any step change when TDH or entrainment rates are plotted against gas velocity. This raises the question as to whether turbulent fluidization exists and, even if it does, whether it matters, since bed expansion, heat transfer coefficients and chemical conversion also show only gradual changes over the range of gas velocities at which turbulent fluidization is believed to start.

CONCLUSIONS

In this work, it was found that as gas velocity was increased up to near-transport conditions (1.2 m/s for this FCC), changes in TDH, bed expansion, and solids concentration in the freeboard all occurred gradually giving no indication that any change in flow regime had occurred.

When used in conjunction with bubble velocities at the bed surface based on equilibrium bubble sizes, several correlations. for TDH give reasonable agreement with experiments carried out at ambient conditions in a 290mm diameter column. Use of the predicted equilibrium bubble size [Equation 1] and the Zenz [12] graphical correlation for TDH gives the best agreement. Sparse data available in the literature suggests that large commercial units operating at temperature and pressure require larger transport disengagement heights.

NOTATION

a	exponent in Wen-Chen equation (-)
A	cross-sectional area of column (m^2)
d_b	equivalent volume bubble diameter (m)
d_{Bo}	equivalent volume bubble diameter at the bed surface (m)
d_{sv}	surface-volume mean size of powder (m)
D	diameter of column (m)
E_o	entrainment flux at surface of bed (kg/m^{2}s)
$E\alpha$	entrainment flux above TDH (kg/m^{2}s)
g	acceleration due to gravity (m/s^2)
K	parameter in Equation (10) (-)
S	Stokes number (-)
u'_o	fluctuation in freeboard gas velocity (m/s)
U	superficial velocity of gas (m/s)
U_{mb}	minimum bubbling velocity (m/s)
U_{mf}	minimum fluidization velocity (m/s)
U_{bs}	bubble rise velocity (m/s)
V_t	particle terminal velocity (m/s)
V'_t	terminal velocity of a particle 2.7 times average size (m/s)
β	turbulence decay constant (-)
ϕ	bubble velocity multiplier (Equation 13)
μ	viscosity of gas (kg/ms)
ρ	density of gas (kg/m^3)
ρ_p	density of particle (kg/m3)

LITERATURE CITED

1. Xie, H.Y., Ph.D. Dissertation, University of Bradford, (1993)

2. Rhodes, M.J. and D.Geldart, .In "Fluidization V", K.Ostergaard and A.S.Sorensen, (Eds.) Engineering Foundn., New York. p.281, (1986)

3. Brereton, C.M.H. and J.R.Grace, *Chem.Eng. Res.& Des.*(I.Chem.E. Lond.),70, 246, (1992)

4. Newton, D. In Proceedings of First International Particle Technology Forum, Denver, Colorado, Aug.17-19 1994, A.I.Ch.E., New York, Vol.1, p.409

5. Li, Y. and M. Kwauk, In "Fluidization III" J.R.Grace and J.M.Matsen (Eds.) Plenum Press, New York p.537,(1980)

6. Sciazko, M., J.Bandrowski, and J.Raczek, *Powder Technol.* 66, 33, (1991)

7. Johnsson, F., A.Svensson and B.Leckner, In "Fluidization VII" O.Potter and D.J.Nicklin, Engineering Foundn., New York, p.471,(1992)

8. Geldart, D., N.Broodryk, and A.Kerdoncuff, *Powder Technol.* 76, 175, (1993)

9. Chan, I.H. and T.M.Knowlton, Paper presented at Ann. Meeting of A.I.Ch.E.,(1984)

10. Schuurmans, H.J.A. 8th Intl. Symp. on Chem. React. Eng., I.Chem.E. Symp. Ser. No.87, 495, (1984)

11. Fournol, A.B., M.A.Bergougnou, and C.G.J.Baker, *Can.J.Chem.Eng.*,51,401,(1973)

12. Zenz, F.A. *Chemical Engineering,* 90, No.24, 61, (1983)

13. Geldart, D., Proc. of Short Course on Gas Fluidization, University of Bradford, (1976)

14. Geldart, D. and H.Y.Xie, In "Fluidization VII" O.Potter and D.J.Nicklin (Eds.) Engineering Foundn., New York, p.749,(1992)

15. Hatano, H. and M.Ishida *Powder Technol.* 35, 201, (1983)

16. Wen, C.Y. and L.H.Chen, *A.I.Ch.E J.*,28, 117, (1982)

17. Horio, M., A.Taki, Y.S.Hsieh, and I.Muchi. In "Fluidization III" J.R. Grace and J.M. Matsen (Eds.) Plenum Press, New York, p.509, (1980)

18. Baron, T., C.L.Briens, and M.A.Bergougnou, *Can.J.Chem.Eng.* **66**,749, (1988)

19. Amitin, A.V., Martyushin, I.G., and Gurevich, D.A. *Khim.Tk.Top.Mas.,***3**,20, (1968)

20. Pemberton, S.J. and J.F.Davidson, *Chem.Eng.Sci.* **41**, 253, (1986)

21. Frantz, J.F. and W.G.Juhl, Paper 42b, 71st National Meeting, Dallas, Feb.20-23, 1972, A.I.Ch.E., New York
22. Wolfarth, A. Intl. Symp.in Two Phase Systems, Technion, Haifa, Israel,(1971)

23. Zenz, F.A. and N.A.Weil, *A.I.Ch.E. Jl.* **4**,472, (1958)

24. Werther, J. In "Fluidization IV" D. Kunii and R.Toei (Eds.) Engineering Foundn., New York, p.93, (1983)

25. Geldart, D. and D.J.Pope , *Powder Technol.* **34**, 95,(1983)

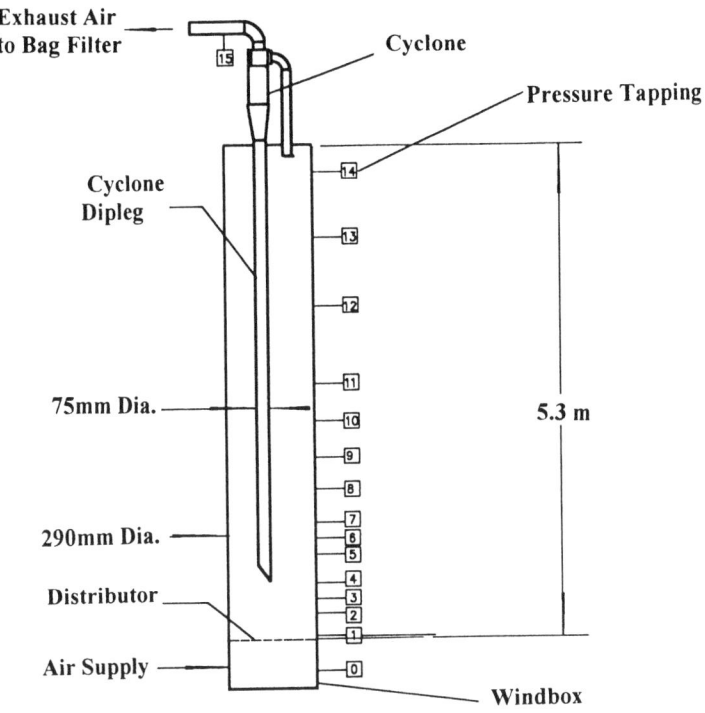

Figure 1 Diagram of fluidization column

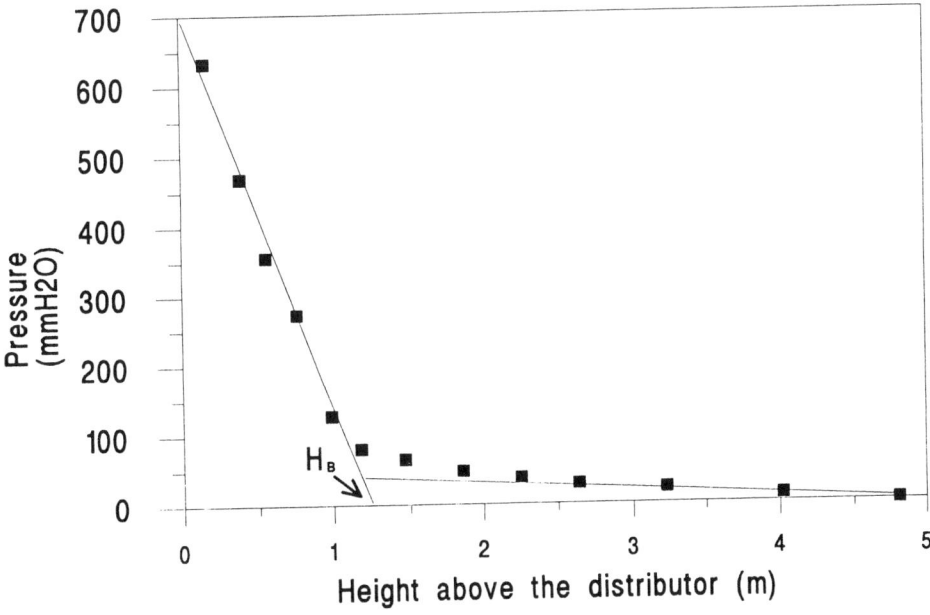

Figure 2 Pressure distribution in the column at U = 0.65 m/s

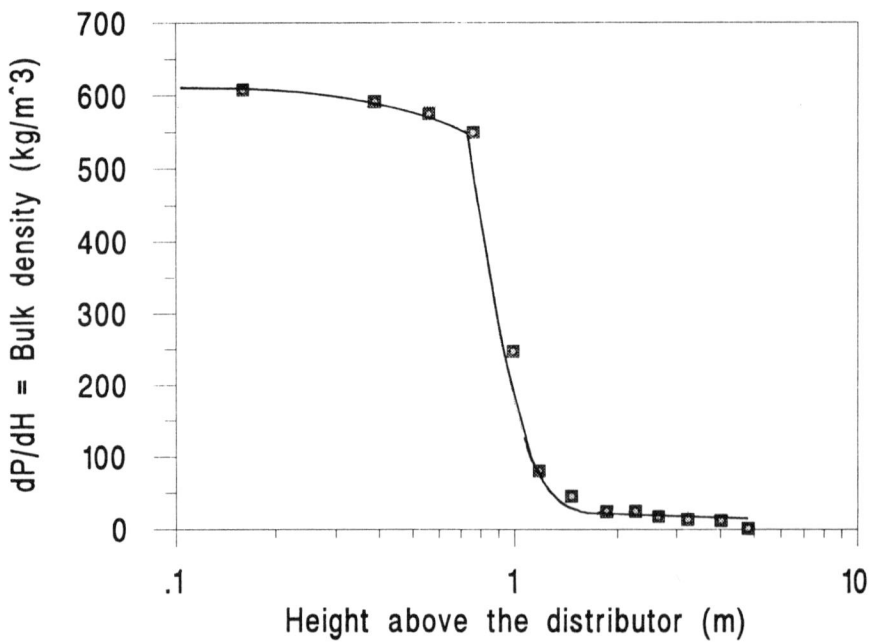

Figure 3 Pressure gradient in the column at U = 0.65 m/s

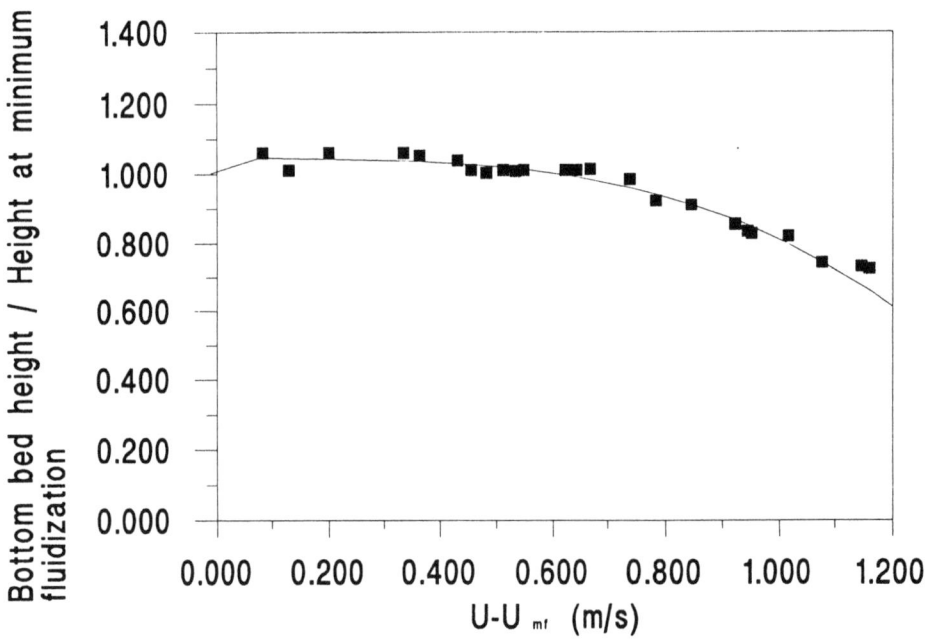

Figure 4 Dimensionless height of bottom bed vs. $U-U_{mf}$

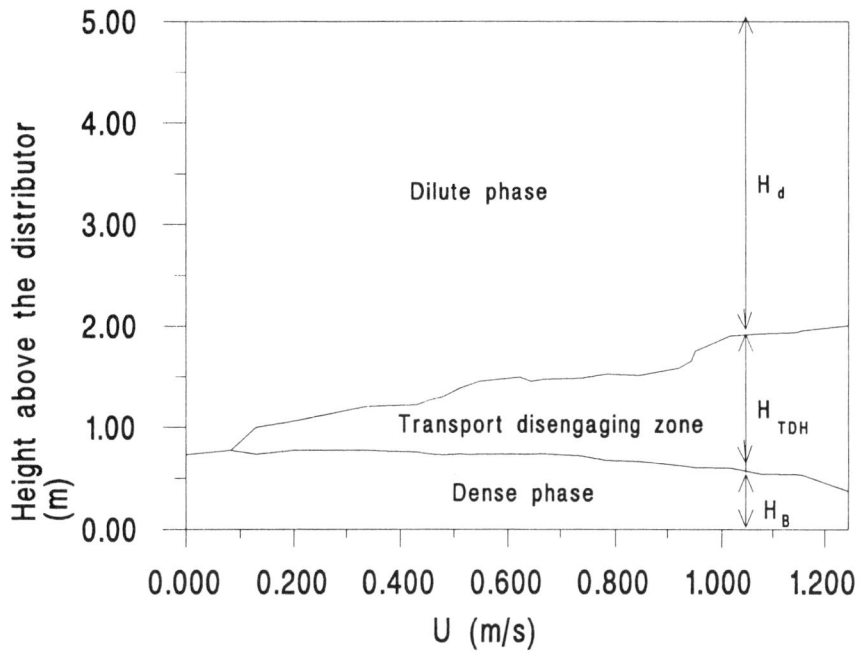

Figure 5　Heights of zones in column vs. gas velocity

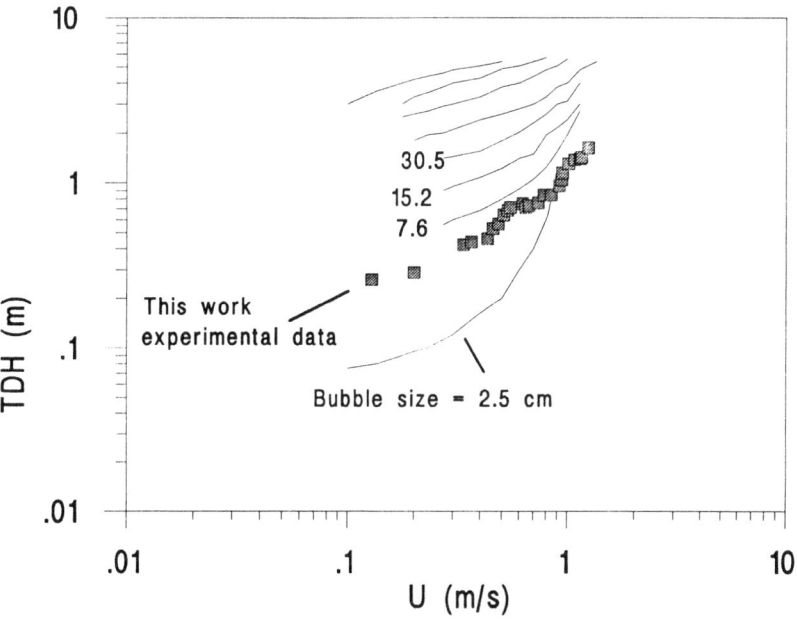

Figure 6　Experimental data compared with Zenz's graphical correlation

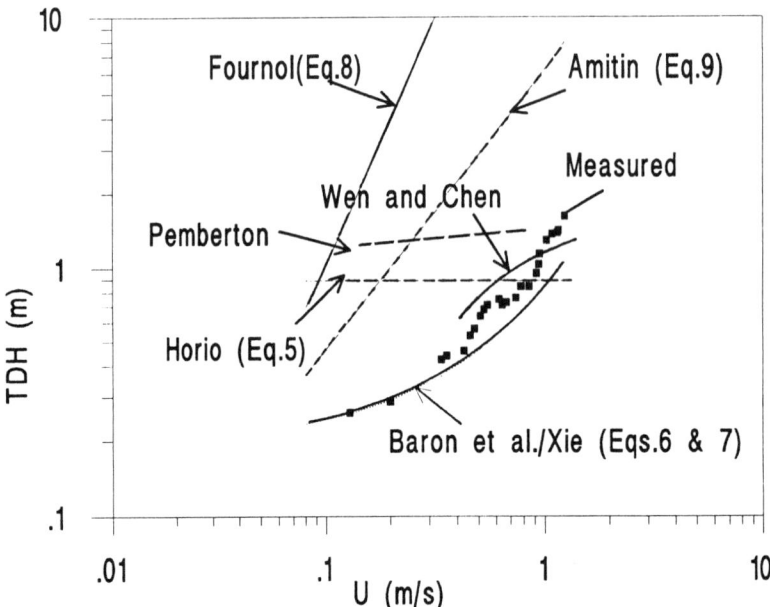

Figure 7 Experimental data compared with predictions from various correlations

Dimension Measurements of Hydrodynamic Attractors in Circulating Fluidized Beds

Lu Huilin and D. Gidaspow
Illinois Institute of Technology, 10 West 33rd Street, Chicago, IL 60616

J.X. Bouillard
Argonne National Laboratory, 9700 South Cass Avenue, Argonne, IL 60439

Experimental measurements of correlation dimension, Kolmogorov entropy and Lyapunov exponent of CFB chaotic attractors were performed from differential pressure and γ-Ray porosity time series recorded along the height of a cold experimental circulating fluidized bed operating with 75-micron particle diameter FCC catalyst. These measurements showed an attractor dimension invariance with respect to the type of measurement performed. Pressure differential and γ-Ray measurements show the existence of a low order hydrodynamic attractor whose dimension varies between 1.5 and 2.0 over the range of gas velocities and solids fluxes studied. Differential pressure measurements indicate that the attractor dimension slightly lowers in the denser and lower section of the CFB and decreases in a similar fashion at higher solids fluxes. Radial γ-Ray porosity measurements indicate that the attractor dimension does not significantly vary across the bed cross section but shows a tendency to slightly lower near the wall region of the riser.

Introduction

Deterministic chaos theory [1, 2, 3, 4, 5, 6, 7, 8, 9, 10, 11] provides, at least in principle, the necessary tools to analyze the chaotic behavior of hydrodynamic systems and to estimate the minimum degrees of freedom necessary to describe such systems. To describe the hydrodynamics of circulating fluidized beds, computer modelers have often had recourse to a fundamental approach which involves the numerical solution of the three-dimensional Navier-Stokes multiphase equations [12, 13, 14, 15, 16, 17, 18, 19, 20, 21, 22, 23, 24, 25, 26]. This approach is often expensive in computer resources (time, computer capacity, computer speed) and is often not readily applicable for routine design and/or control of fluidized bed combustor (FBC) industrial units. However, such modeling complexity could be mitigated and hopefully streamlined, should lower degrees of freedom be found experimentally. The deterministic chaos theory allows the measurement of the degrees of freedom by examining the experimental time series of system variables of interest. Such an approach was recently taken by Bouillard and Lyczkowski [27] who showed that for the case of slugging beds, the hydrodynamic behavior of the bed should be described by no more than two or three independent system variables. In this study, a similar approach is undertaken by independently analyzing the time series of experimentally measured differential pressure and void fraction fluctuations along the riser height of a circulating fluidized bed (CFB).

In the following sections, the experimental facility is described. Major frequencies of pressure and void fraction fluctuations were identified using spectral frequency analysis. Poincare maps of the attractor were performed and correlation dimensions measured. The hydrodynamic behavior of the CFB appears to be chaotic of low dimension. This dimension is shown to be *independent* of the type of measurement performed (Pressure or γ-Ray). A discussion on the theoretical implications of this finding is presented.

Experimental

A cold-model circulating fluidized bed (CFB) test unit, constructed at the Illinois Institute of Technology (IIT) and shown in Figure 1, was used to obtain detailed pressure fluctuation data along the height of the riser. Air enters the CFB in its bottom U section with 75 μm FCC catalyst particles and flows through a distributor (38.1 mm ID) into a

7.5 cm ID clear acrylic riser. This riser is approximately 6.6 m long and is connected at its top to a smoothly bended exit PVC pipe to limit exit effects. The effluent stream enters a primary cyclone where separation of gas and particles takes place. Additional separation is obtained by a smaller and more efficient secondary cyclone. To obtain optimal cyclone efficiency and to insure an adequate downflow of solids, the solids discharge into a fluidized-bed receptacle. The cyclone dipleg is carefully positioned at the proper depth in this fluidized-bed receptacle to provide adequate static head for sufficient solids downflow. Solids overflow from the fluidized-bed receptacle into a storage hopper which consists of a 10 cm ID PVC pipe and is equipped at its bottom with a slide valve to regulate the solids flow and loading in the riser (See Figure 1). Several experiments were performed where the fluidizing velocity and solids flux were varied. More information regarding the experimental setup of this CFB can be found in Miller and Gidaspow [28].

γ-Ray densitometry is used to measure local instantaneous porosity. A 500 mCi Cs-137 source, of 30-year half-life, emitting γ rays of 0.66 MeV was beamed upon the gas solids mixture flowing in the riser. The detector is a NaI (Ti) crystal scintillation detector (Teledyne, ISOTOPES/S-44-I/2). It consisted of a 2 mm thick, 5.08 cm diameter tube with a 0.13 mm thick Berylium window. The transmitted radiation was converted to electrical pulses by photo-multiplier (Model 266, EG&G Ortec). The converted electrical signals were passed through a series of data conditioners including a preamplifier (Model 113, EG&G), an amplifier and a double channel analyzer (DCA) (Model 778). The DCA has been used to remove low energy level noises by screening only inputs that display energy above a threshold level. Bed porosity measurements are based on the Beer-Lamber law, which states:

$$\frac{I}{I_o} = e^{-K\rho L} \qquad (1)$$

where I is the intensity of transmitted radiation, I_o is the intensity of incident radiation, K is an attenuation coefficient, $\rho = \rho_s \epsilon_s$ is the bulk density of the material to be measured and L is the radiation path length. From Equation 1, ρ is measured and depends upon the solids volume fraction of the material. The attenuation constant for the cracking catalyst used in this study was measured to be $K = 0.0248$ m^2/Kg and the solids specific density ρ_s was 1560 Kg/m^3. Using this technique, γ-Ray measurements can be made a very reliable tool to measure transient solids volume fraction as discussed by Gidaspow and Huilin [29, 30]. Differential pressure transducers manufactured by Validyne Corporation were used to measure pressure drops along the riser height at two locations identified as upper (locations 1-2) and lower (locations 2-3) positions indicated in Figure 1 (Bouillard and Miller [31]). γ-Ray measurements were made at a lower and denser part of the CFB riser, at about 1.8 m from the bottom of the riser and at a leaner and upper region at about 4.4 m from the bottom of the riser.

Deterministic Chaos Analysis

Differential pressure and void fraction fluctuations were recorded along the CFB height and across the CFB horizontal section as shown in Figures 2 and 3. Data acquisition sampling frequency was set to 40 Hz. In order to eliminate spurious effects of initial conditions on data analysis, data taken over the first second were systematically eliminated from time-series data sets. From time series of experimentally measured differential pressure and void fraction fluctuations, the correlation integral, $C(\epsilon)$, is calculated as follows:

$$C(\epsilon) = \frac{1}{N^2}[\text{number of pairs } (i,j), |x_i - x_j| < \epsilon] \qquad (2)$$

where x_i is the m dimensional reconstruction vector defined as:

$$x_i = [p(i), p(i+\tau), p(i+2\tau), ..., p(i+(m-1)\tau)] \qquad (3)$$

where $p(i)$ is the measured pressure or void fraction fluctuation, τ is the time delay for the attractor reconstruction and m is the embedding dimension. Grassberger and Procaccia showed that the information correlation integral $C(\epsilon)$ scales as ϵ^D where D is the correlation dimension of the attractor [1]. The Lyapunov exponent measures the rate at which two adjacent trajectories on the attractor diverge. If two trajectories are distant from one another by $d(o)$, the distance between these two trajectories after the time t is given by:

$$d(t) = d(o)2^{Kt} \qquad (4)$$

where K is the Lyapunov exponent. Given a time series, a Lyapunov exponent can be estimated as follows:

$$K = \frac{1}{t_n - t_o} \sum log_2 \frac{d(t_i)}{d(t_{i-1})} \qquad (5)$$

The criterion for existence of chaos is given by:

- $K > 0$ the system is chaotic (i.e. the trajectories tend to diverge)
- $K < 0$ the system is regular (i.e. the trajectories tend to converge)

In this study, the Lyapunov exponent was found to be positive, indicating the existence of a chaotic attractor.

The Kolmogorov entropy is directly related to the positive Lyapunov exponents of chaotic systems, and is therefore a direct measure of chaos. The Kolmogorov entropy represents the sum of the positive Lyapunov exponents, and is therefore dominated by the largest positive exponent. The Kolmogorov entropy is characteristic of the rate of generation of information of a system. It can be estimated by considering the fraction of pairs separated by a distance smaller than a given r_o, in an embedded phase space of dimension m. This fraction can be expressed as follows:

$$C(r_o, m) \sim exp(-K_2 m\tau) \qquad (6)$$

where τ is the time delay. For a periodic system, the Kolmogorov entropy K_2 is zero, indicating that no information is generated as the system evolves. The system is said to be predictable at all times. A purely random system shows an infinite Kolmogorov entropy and is therefore not predictable. For a deterministic chaotic system, the Kolmogorov entropy is finite and positive.

Results and Discussions

For various operating conditions, the correlation dimensions were measured by embedding the signal in a phase space dimension which varied between 5 and 8. Correlation integrals were measured from the pressure fluctuation data as shown in Figures 4, 5 and 6. Typical low order attractors were found with a fractal dimension which ranged between 1.5 and 1.9. In a similar data analysis, γ-Ray density measurements were analyzed as indicated in Figures 7 and 8. As can be noticed, the attractor dimension essentially converges with an embedding dimension varying between 5 to 7. γ-Ray porosity measurements reveal a low order attractor whose dimension also varies between 1.5 and 2. This low order attractor dimension is further corroborated by γ-Ray density measurements carried out over the radial riser distance as shown in Figures 8.a and 8.b. From these figures, the attractor dimension is seen to slightly decrease near the wall of the riser where the solids concentration is greater. This dependency of the attractor dimension with the solids loading corroborate with the increase of the attractor dimension at more dilute regimes (i.e. at lower solids flux) obtained from differential pressure fluctuation data as displayed in Figures 4 and 6.

Hence the bed characteristic becomes less "chaotic" as the solids loading increases. This trend reinforces the view that the presence of fine solids in a gas stream reduces the amount of "turbulence" and therefore the complexity of the flow.

The Kolmogorov entropy was calculated from the experimental porosities measured by γ-Ray densitometry. A Kolmogorov entropy convergence study on the embedding dimension and the time delay was performed and indicated that results converged for an embedding dimension ranging from 5 to 7. The variations of the Kolmogorov entropies with respect to the riser radial distance are shown in Figures 9 a and 9 b. The Kolmogorov entropy varies between 7 to 12 bits/second with minor variations across the CFB cross section. Entropy data suggest that the accurate temporal and spatial prediction of CFB hydrodynamics as an initial value problem would be difficult due to the important loss of information beyond a second of real time. This result may have strong repercussions on the development of mechanistic computer models which are used to describe circulating fluidized bed hydrodynamics as initial-value problems. Even though the accurate hydrodynamic computations of CFBs may not be possible, a temporal and spatial hydrodynamic envelope for CFB may conceivably be accurately predicted and be useful for engineering design purposes. More research, however, needs to be performed in this area to ensure reliability and meaningfulness of computer predictions based on multi-phase flow Navier-Stokes equations over long periods of time.

Although, pressure fluctuations and γ-Ray measurements were made at different locations and have different spatial and accuracy resolutions, they unequivocally indicate that our CFB exhibits a low order attractor of dimension smaller than 2. This result proves that the CFB Hydrodynamics is governed by two independent variables, which could conveniently chosen to be the void fraction and the gas/solids relative velocity. This result also indicates that the description of such system entails at least three independent partial differential equations. This is usually the case for the resolution of the void fraction, the gas and the solids velocity variables which are expressed in terms of the phase continuity, and the gas and solids momentum equations. Note also that an upper limit to the number of variables required to describe the dynamics of the system may be given by n = 2d + 1, where d is the attractor dimension or n=5 in our case. This implies that at most n or 5 ordinary differential equations are needed to describe this system. This number is also the minimal embedding dimension, m , that was used to calculate the attractor dimension.

For practical design applications, this type of analysis provides a sound basis for determining how many important variables should enter into a design correlation. For small diameter CFBs, as the one used here, this study suggests that hydrodynamic correlations should involve at least two independent variables: the bed porosity and the relative velocity ($v_g - v_s$). This is indeed the case for Ergun's equation, for which $\Delta P/\Delta L$ is expressed essentially in terms of the bed porosity and the relative velocity [32]. By the same token, this study stresses also the practical importance of the drift flux theory, which has proved successful for the design of CFB risers and standpipes [33]. In the drift flux theory, only two major variables, namely the bed porosity and the relative velocity, need to be considered as recently reviewed by Gidaspow [32]. Hence, this study shows that deterministic chaos theory can indeed be used to determine the type and amount of experimental data one should take to develop useful design correlations. Such application of the theory has not yet been reduce into practice in the development of scaling laws. However, one can foresee the potential cost-savings benefits of this theory by reduction of experimental data base requirements.

For larger diameter fluidized beds, where cross sectional flow phenomena may become significant, flow characteristics should be expressed in terms of the bed porosity and the three spatial components of the relative velocity. This would lead to a hydrodynamic attractor dimension of at least 3 or 4 depending on the system investigated. We should also recall that the hydrodynamic modeling of fluidization entails at least the solution of the bed porosity, the pressure, and the solids and gas velocities which globally yields four hydrodynamic variables in one dimension, six in two dimensions, and eight in three dimensions. It is interesting to observe that the natural coupling between the gas and solids phase flow fields through constitutive equations indeed reduces the complexity of the problem to lower dimensions.

Conclusion

A low hydrodynamic attractor dimension in a CFB riser was found from two types of measurement, namely pressure differentials and γ-ray densitometry. A γ-Ray source strength of 500 m Ci was sufficient to accurately capture the chaotic dynamics of the CFB. The attractor dimension which neighbors 2 suggests that two main independent variables can describe the system. These two fundamental variables can be conveniently chosen to be the porosity and the relative velocity. The Kolomogorov entropy production ranged from 7 to 11 bits per second. Such high rates indicate a strong rate of information loss over one second of real time. This result is indicative of potential underlying difficulties in the development of accurate predictions of the temporal and spatial hydrodynamic behavior of CFBs using initial-value models. Though precise hydrodynamic information may inevitably be lost by the use of computer models, it is believed that a temporal and spatial hydrodynamic envelope can still be accurately predicted and be meaningful for routine design purposes.

Acknowledgments

This work was supported by the US Department of Energy under the contract W-31-109-Eng-38 and the National Science Foundation, Grant No CTS-9305850. We thank Dr. R. W. Lyczkowski of Argonne National Laboratory for his enjoyable technical comments and insights.

Bibliography

[1] Grassberger P. and I. Proccacia. Measuring the strangeness of strange attractors. *Physica*, 9D:189, (1983).

[2] Grassberger P., Th. Schreiber, and C. Schraffrath. Non linear time sequence analysis. *International of Bifurcation and Chaos*, 1(3):521–547, (1991).

[3] F. Takens. *Lecture Notes in Mathematics*, chapter Detecting Strange Attractors in Turbulence, page 366. Volume 898, Springer and Verlag, (1981).

[4] Schouten J. C. and C. M. Van Den Bleek. Chaotic behavior in a hydrodynamic model of a fluidized bed reactor. In *Proc. of the 11 International Conference on Fluidized Bed Combustion*, pages 459–466, E. J. Anthony and Canmet, ASME, New York, Montreal, April 21-24, 1991.

[5] Schouten J. C., M. L. M. Van Der Stappen, and C.M. Van Den Bleek. Deterministic chaos analysis of gas-solids fluidization. *AIChE Annual Meeting*, 1991.

[6] Schouten J. C. and C. M. Van Der Bleek. Chaotic hydrodynamics of fluidization: consequences for scaling and modeling of fluidized bed reactors. *AIChE, Annual Meeting*, (1991).

[7] Van Der Stappen M. L. M., J. C. Schouten, and C. M. Van den Bleek. Application of deterministic chaos in understanding the fluid dynamic behavior of gas-solids fluidization. *AIChE Annual Meeting, Miami*, (1992).

[8] Daw C. S. and J. S. Halow. Characterization of voidage and pressure signals from fluidized beds using deterministic chaos theory. In *Proc. of the 11 International Conference on Fluidized Bed Combustion*, pages 777–786, E. J. Anthony, ASME, New York, Montreal, April 21-24, 1991.

[9] Daw C. S. and J. S. Halow. Evaluation and control of fluidization quality through chaotic time series of pressure drop measurements. *AIChE Annual Meeting, Miami*, (1992).

[10] Daw C. S., T. J. O'Brien, M. Syamlal, and P. Nicoletti. Correlating pressure-drop and voidage measurements with gas-solids contacting in gas-fluidized beds. *AIChE Annual Meeting, Miami*, (1992).

[11] Daw C. S. and J. S. Halow. Modeling deterministic chaos in gas-fluidized beds. *AIChE Symposium Series*, 88(289):61–69, (1992).

[12] Gidaspow D. Hydrodynamics of fluidization and heat transfer: supercomputer modeling. *Applied Mechanics Review*, 39(1):1–23, 1986.

[13] Lun C.K.K., S.B. Savage, D.J. Jeffrey, and N. Chepurniy. Kinetic theories for granular flow: inelastic particles in couette flow and slightly inelastic particles in a general flow field. *J. Fluid Mech.*, 140:223–256, 1984.

[14] Bouillard J.X., R.W. Lyczkowski, and D. Gidaspow. Porosity distributions in a fluidized-bed with an immersed obstacle. *A.I.Ch.E. Journal*, 35(6):908–922, 1989.

[15] Bouillard J.X., R.W. Lyczkowski, S. Folga, D. Gidaspow, and G.F. Berry. Hydrodynamics of erosion of heat exchanger tubes in fluidized bed combustors. *Can. J. Chem. Eng.*, 67:218–229, 1989.

[16] Savage S.B. *Granular Flow at High Shear Rates*, pages 339–357. Academic Press, New York, 1982.

[17] Savage S. B. Streaming motions in a bed of vibrationally fluidized dry granular material. *J. Fluid Mech.*, 194:457, (1988).

[18] Savage S. B. and D. J. Jeffrey. The stress tensor in a granular flow at high shear rates. *J. Fluid Mech.*, 110:225, (1981).

[19] Savage S. B. and M. Sayed. Stresses developed by dry cohesionless granular material sheared in an annular shear cell. *J. Fluid Mech.*, 142:391, (1984).

[20] Savage S. B. Instability of unbounded uniform granular shear flow. *J. Fluid Mech.*, 241:109–123, (1992).

[21] Jenkins J. T. and S. B. Savage. A theory for the rapid flow of identical smooth nearly elastic spherical particles. *J. Fluid Mech.*, 130:187, (1983).

[22] Ding J. and D. Gidaspow. A bubbling fluidization model using kinetic theory of granular flow. *A.I.Ch.E. Journal*, 36(4):523–538, (1990).

[23] Sinclair J. L. and R. Jackson. Gas-particles in a vertical pipe with particle-particle interactions. *A.I.Ch.E. Journal*, 35(9):1473–1486, (1989).

[24] Pita J. P. and S. Sundaresan. Gas-solid flow in vertical tubes. *A.I.Ch.E. Journal*, 37(7):1009–1018, (1991).

[25] Nott P. and R. Jackson. Frictional-collisional equations of motion for granular materials and their application to flow in aerated chutes. *J. Fluid Mech.*, 241:125–144, (1992).

[26] Anderson K. G. and R. Jackson. A comparison of the solutions of some proposed equations of motion of granular materials for fully developed flow down inclined planes. *J. Fluid Mech.*, 244:145–168, (1992).

[27] Bouillard J. X. and R. W. Lyczkowski. Chaotic nature of gas/solids fluidization. *AIChE Annual Meeting, Miami*, (1992).

[28] Miller A. and D. Gidaspow. Dense, vertical gas-solid flow in a pipe. *A.I.Ch.E. Journal*, 38(11):1801–1815, (1992).

[29] Gidaspow D. and L. Huilin. *Understanding the Fluid Dynamic Behavior of Circulating Fluidized Bed Through Chaotic Time Series Analysis*. Technical Report, IIT, (1994).

[30] Gidaspow D. and L. Huilin. *Porosity Fluctuation Investigation in the Circulating Fluidized Bed*. Technical Report, IIT, (1994).

[31] Bouillard and A. L. Miller. Experimental investigations of chaotic hydrodynamic attractors in circulating fluidized beds. *J. Powder Technology*, (1994).

[32] D. Gidaspow. *Multiphase flow and FLuidization: Continuum and Kinetic Theory Descriptions*. Academic Press, 1994.

[33] J. M. Matsen. The rise and fall of recurrent particles: hydrodynamics of circulation. *Fluidization Technology*, 3–11, 1988.

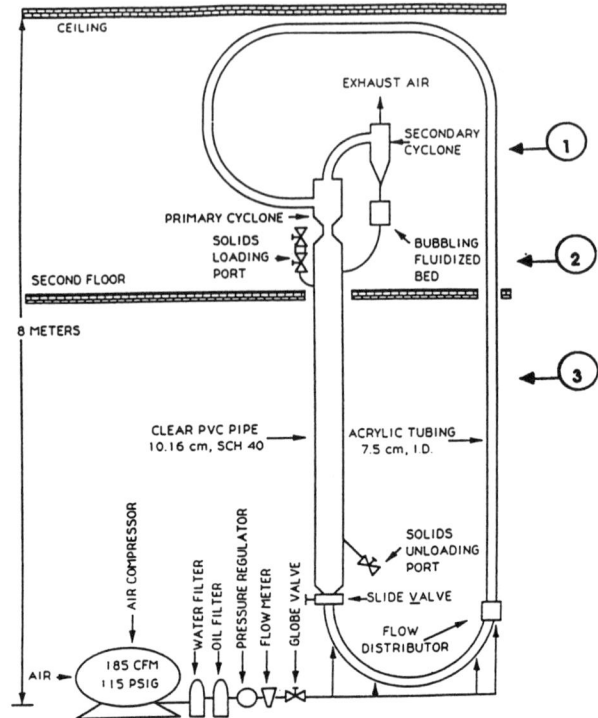

Figure 1: Schematic Diagram of IIT Circulating Fluidized Bed

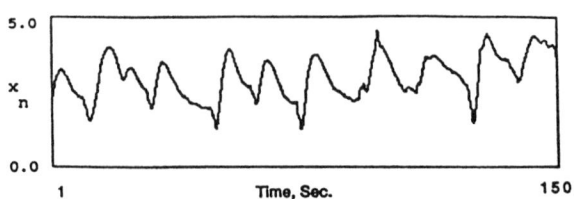

Figure 2: Typical Differential Pressure Fluctuations for Dense Phase Flows of FCC, Solids Flux = 33 Kg/(m^{2}s), Superficial Gas velocity = 2.9 m/s, Upper Position

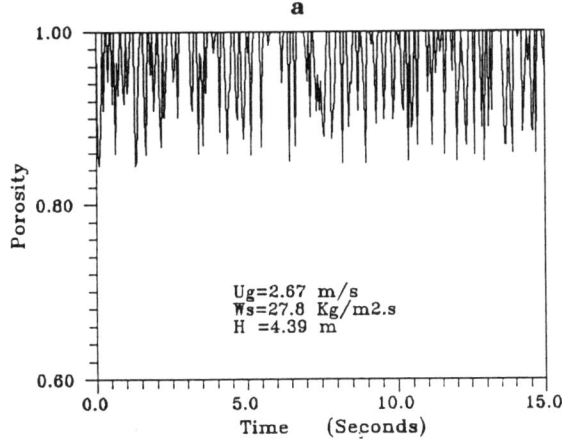

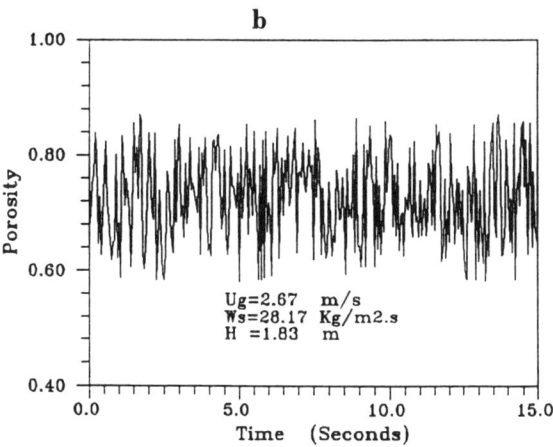

Figure 3: Typical Void Fraction Fluctuation Measured by γ-Ray Densitometry Solids Flux =28 Kg/m²s, Superficial Gas Velocity = 2.7 m/s, a) In the Dilute and Upper Section of the Riser, b) In the Dense and Lower Section of the Riser

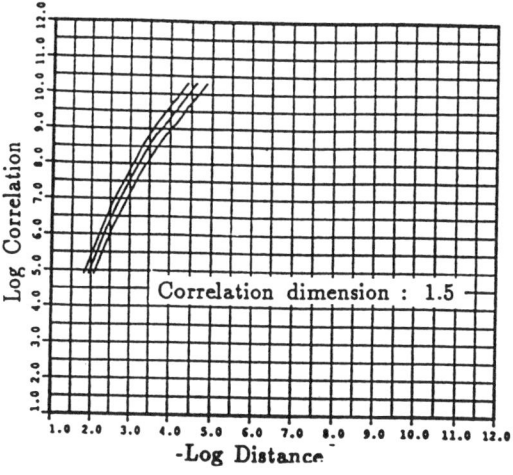

Figure 4: Chaotic Attractor Dimension for Three Embedding Dimensions (From Pressure measurements), (From Pressure Fluctuation Measurements) Superficial Gas Velocity= 2.89 m/s, Solids Flux = 33 Kg/m²s, Upper Position

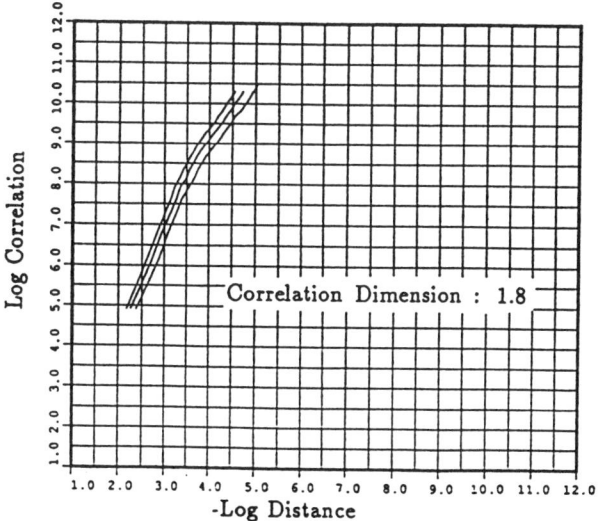

Figure 5: Chaotic Attractor Dimension for Three Embedding Dimensions (From Pressure Measurements), Superficial Gas Velocity= 2.61 m/s, Solids Flux = 20.4 Kg/m²s, Lower Position

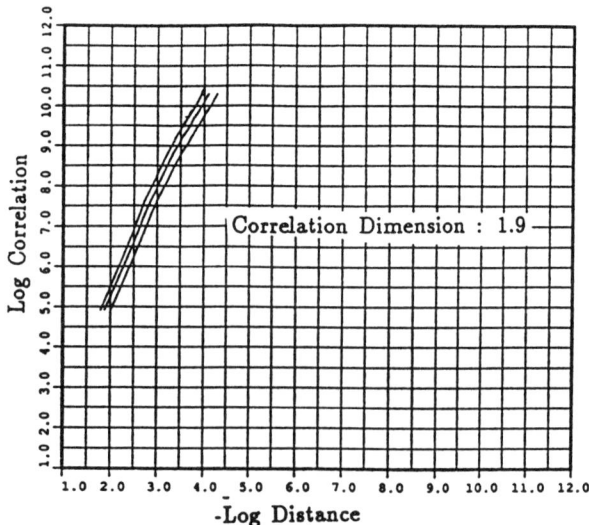

Figure 6: Chaotic Attractor Dimension for Three Embedding Dimensions (From Pressure Measurements), Superficial Gas Velocity= 2.89 m/s, Solids Flux = 20.4 Kg/m²s, Lower Position

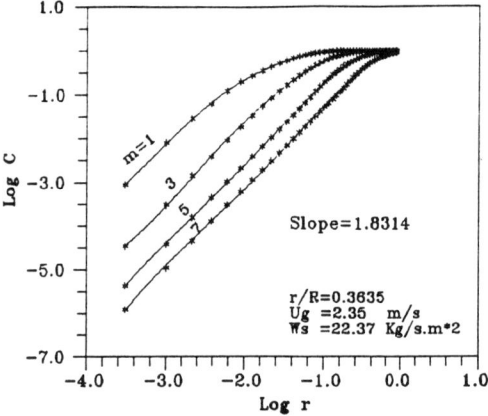

Figure 7: Chaotic Attractor Dimension for Four Embedding Dimensions, (From γ-Ray Fluctuation Measurements) Superficial Gas Velocity= 2.35 m/s, Solids Flux = 22.4 Kg/m²s, Lower Position

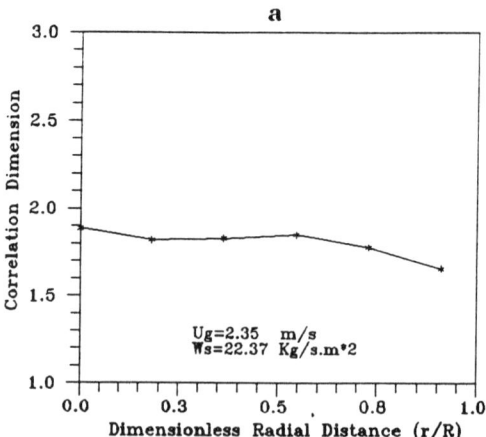

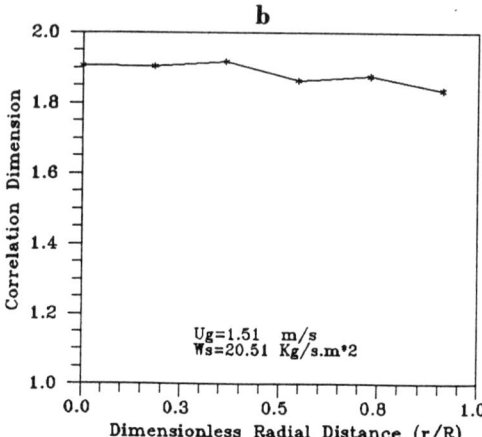

Figure 8: Radial Dependency of the Correlation Dimension Across the CFB Bed section (From γ-Ray Fluctuation Measurements) : a) Superficial Gas velocity:2.35 m/s , solids Flux: 22.4 Kg/sm² and b) Superficial Gas Velocity: 1.51 m/s and Solids Flux 20.5 Kg/sm²

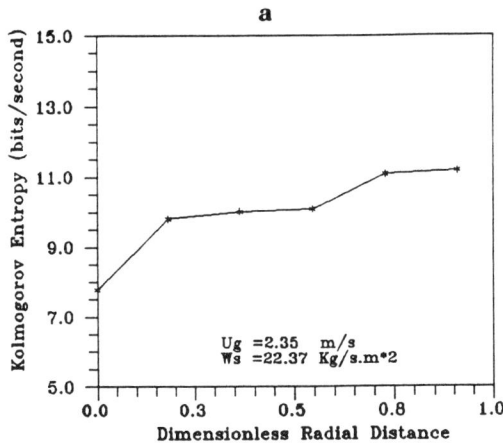

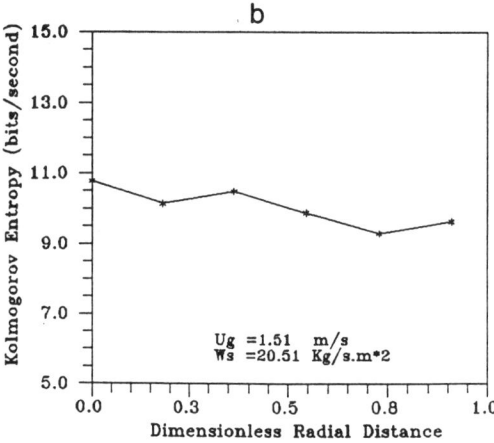

Figure 9: Kolmogorov Entropy Dependency with the CFB Dimensionless Radius of the CFB (From γ-Ray Fluctuation Measurements) : a) Superficial Gas Velocity : 2.35 m/s, Solids Flux: 22.4 Kg/s.m^2 and b) Superficial Gas Velocity: 1.51 m/s and Solids Flux: 20.5 Kg/s.m^2

Gravity Flow of a Fluid-Particle Mixture in a Channel

R. Gudhe
Viscoustech, Inc., 5272 Butler Street, Pittsburgh, PA 15201

K.R. Rajagopal
Department of Mechanical Engineering, University of Pittsburgh, Pittsburgh, PA 15236

M. Massoudi
U.S. Department of Energy, Pittsburgh Technology Center, P.O. Box 10940, Pittsburgh, PA 15236

R.C. Yalamanchili
Department of Mechanical Engineering, University of Pittsburgh, Pittsburgh, PA 15261

Multiphase flows have attained considerable importance in many industrial applications. Flowing mixtures consisting of solid particles entrained in a fluid are relevant to a variety of applications such as pneumatic and hydraulic transport of solid particles and fluidized beds. The governing equations for the flow of fluid-solid mixture in a vertical channel are derived, assuming that the fluid behaves as a linearly viscous fluid and for solid particles we use the constitutive equation proposed by Rajagopal and Massoudi (1990). These equations reduce to a system of coupled non-linear ordinary differential equations. The resulting boundary value problem is solved numerically and the effect of various non-dimensional parameters is studied.

INTRODUCTION

In recent years there has been considerable interest in understanding the behavior of flowing granulate materials. The situations considered include handling of such substances as coal, agricultural products, and other particulate solids, and more complicated processes such as fluidization, combustion, avalanches etc. [cf. Gudhe et al. (1993)].

In earlier studies, Johnson et al. (1991 a,b) formulated a two-phase flow theory based on the theory of interacting continua (or mixture theory) [cf. Bowen (1976)]. They presented numerical solutions for two boundary value problems. The mixture theory, is in a sense, a homogenization approach in which each component is regarded as a single continuum and at each instant of time, every point in space is considered to be occupied by a particle belonging to each component of the mixture [cf. Massoudi (1986)].

Balance laws are then written for each component which takes into account interaction with other constituents.

GOVERNING EQUATIONS

The balance laws for a purely mechanical system (no thermal effects or chemical reactions) are the conservation of mass and conservation of linear momentum in a fluid and solid are

$$\frac{\partial \rho_1}{\partial t} + \text{div}(\rho_1 v_1) = 0, \tag{1}$$

$$\frac{\partial \rho_2}{\partial t} + \text{div}(\rho_2 v_2) = 0, \tag{2}$$

and,

$$\rho_1 \frac{D_1 v_1}{Dt} = \text{div} T_1 + \rho_1 b_1 + f_1, \tag{3}$$

$$\rho_2 \frac{D_2 v_2}{Dt} = \text{div} T_2 + \rho_2 b_2 - f_2 \tag{4}$$

where b represents the body force, f_1 represents the mechanical interaction (local exchange of momentum) between the components, ρ_1 and ρ_2 are the bulk densities of the fluid and solid constituents and T_1 and T_2 denote the partial stress tensors of the fluid and solid respectively.

The constitutive relations for the fluid and solid are given by [cf. Rajagopal and Massoudi (1990), Goodman and Cowin (1971, 1972)].

$$\mathbf{T}_f = [-P(\rho_1) + \lambda_f(\rho_1)\text{tr}\mathbf{D}_1]\mathbf{I} + 2\mu_f(\rho_1)\mathbf{D}_1 \tag{5}$$

$$\mathbf{T}_s = [\beta_o(v) + \beta_1(v)\nabla v \bullet \nabla v + \beta_1(v)\text{tr}\mathbf{D}_2]\mathbf{I} + \beta_4(v)\nabla v \otimes \nabla v + \beta_3(v)\mathbf{D}_2 \tag{6}$$

where P is the fluid pressure, λ_f and μ_f are the viscosities, $\mathbf{D}_1$ is the stretching tensor associated with the fluid, v is the volume fraction of the solids, $\mathbf{D}_2$ is the stretching tensor associated with the motion of solids, $\beta_o(v)$ is similar to pressure in a compressible fluid and is given by an equation of state, $\beta_2(v)$ is like the second coefficient of viscosity in a compressible fluid, $\beta_1(v)$ and $\beta_4(v)$ are the material parameters connected with the distribution of the granular materials and $\beta_3(v)$ is the viscosity of the granular materials.

The mechanical interaction between the mixture components, f_1 is written as [cf. Johnson et al. (1991 a)]:

$$\mathbf{f}_1 = A_2 F(v)(V_2 - V_1) + A_3 v (2\text{tr}\mathbf{D}_1^2)^{-\frac{1}{4}} \mathbf{D}_1 (V_2 - V_1) \tag{7}$$

where $F(v)$ is the dependence of the drag coefficient on the volume fraction. The terms in equation (7) reflect the presence of drag and slip-shear lift. Following Johnson et al. (1991 a) we assume $F(v)$ to be of the form given by

$$F(v) = \frac{v(4+3v+3\sqrt{8v-3v^2})}{(2-3v^2)},$$
(8)

Notice that the above equation diverges as $v \to 0$.

A mixture stress tensor is defined as [cf. Green and Naghdi (1969)]:

$$\mathbf{T}_m = \mathbf{T}_1 + \mathbf{T}_2,$$
(9)

where,

$$\mathbf{T}_1 = (1-v)\mathbf{T}_f, \text{ and } \mathbf{T}_2 = v\mathbf{T}_s,$$
(10)

so that the mixture stress tensor reduces to that of a pure fluid as $v \to 0$ and to that of a granular solid as $v \to 1$.

Consider the flow of a fluid-solid mixture down a vertical channel due to the action of gravity [cf. Figure 1]. In this problem, we consider steady one dimensional flow of incompressible granular materials and also the fluid is incompressible. Let X denote the direction of motion and let the vertical plates be located at Y = -1 and Y = 1. The flow field is assumed to be of the form given by

$$v = v(y)$$
$$V_1 = V(y)\mathbf{i} \quad \text{(Fluid)}$$
$$V_2 = u(y)\mathbf{i} \quad \text{(Solid)}$$
(11)

Following Rajagopal and Massoudi (1990), we assume that the material parameters corresponding to the solid to be of the form

$$\beta_o = B_o v$$
$$\beta_1 = B_1(1+v+v^2)$$
$$\beta_2 = B_2(v+v^2)$$
$$\beta_3 = B_3(v+v^2)$$
$$\beta_4 = B_4(1+v+v^2)$$
(12)

With the above flow field the conservation of mass are automatically satisfied. The balance of linear momentum in the non-dimensional form reduces to

$$\overline{y} = \frac{y}{h}, \quad \overline{u} = \frac{u}{u_o}, \quad \overline{V} = \frac{V}{u_o}, \quad \overline{\rho}_1 = 1, \quad \overline{\rho}_2 = \frac{\rho_2}{\rho_1}$$

$$(1-v)\frac{d^3\overline{V}}{d\overline{y}^3} - 2\frac{dv}{d\overline{y}}\frac{d^2V}{d\overline{y}^2} - \frac{d^2v}{d\overline{y}^2}\frac{d\overline{V}}{d\overline{y}} - R_1\frac{dv}{d\overline{y}} +$$

$$D_1\left[F(v)\left(\frac{d\overline{u}}{d\overline{y}} - \frac{d\overline{V}}{d\overline{y}}\right) + \frac{dF}{dv}(\overline{u} - \overline{V})\frac{dv}{d\overline{y}}\right] = 0$$

$$(v+v^2)\frac{d^2V}{d\overline{y}^2} + (1+2v)\frac{dv}{d\overline{y}}\frac{d\overline{u}}{d\overline{y}} + R_2 v -$$
$$D_2 F(v)(\overline{u} - \overline{V}) = 0$$

$$2(1+v+v^2)\frac{dv}{d\overline{y}}\frac{d^2v}{d\overline{y}^2} + (1+2v)\left(\frac{dv}{d\overline{y}}\right)^3 +$$

$$B\frac{dv}{d\overline{y}} + Lv\left|\frac{d\overline{V}}{d\overline{y}}\right|^{-\frac{1}{2}}\frac{d\overline{V}}{d\overline{y}}(\overline{u} - \overline{V}) = 0$$
(13)

with

$$\overline{y} = \frac{y}{h}, \quad \overline{u} = \frac{u}{u_o}, \quad \overline{V} = \frac{V}{u_o}, \quad \overline{\rho}_1 = 1, \quad \overline{\rho}_2 = \frac{\rho_2}{\rho_1}$$
(14)

where,

$$C_2 = \frac{A_2 L}{\rho_1 u_o}; \quad Fr = \frac{u_o^2}{Lg}; \quad Re = \frac{\rho_1 u_o L}{\mu_f};$$

$$\overline{B}_o = \frac{B_o}{\rho_1 u_o^2}; \quad \overline{B}_1 = \frac{B_1}{\rho_1 u_o^2 L^2}$$

$$\overline{B}_3 = \frac{B_3}{\rho_1 u_o L}; \quad \overline{B}_4 = \frac{B_4}{\rho_1 u_o^2 L^2}; \quad C_3 = \frac{A_2 L^{\frac{1}{2}}}{\rho_1 u_o^{\frac{1}{2}}}.$$

(15)

and let us define the following non-dimensional parameters.

$$D_1 = C_2 Re; \quad D_2 = \frac{2C_2}{\overline{B}_3}; \quad R_1 = \frac{Re}{Fr};$$

$$R_2 = \frac{2\overline{\rho}_2}{Fr \overline{B}_3}; \quad B = \frac{\overline{B}_o}{\overline{B}_1 + \overline{B}_4}; \quad L = \frac{-C_3}{\overline{B}_1 + \overline{B}_4}$$

(16)

Equations $(13)_{1,2,3}$ are solved numerically subjected to the following boundary conditions in the non-dimensional form:

$$\overline{u} = 0$$
$$\overline{V} = 0 \quad \text{at} \quad \overline{y} = -1 \quad (17)$$

$$\overline{u} = 0$$
$$\overline{V} = 0 \quad \text{at} \quad \overline{y} = 1 \quad (18)$$

$$N = \int_{-1}^{1} v d\overline{y} \quad (19)$$

$$Q = \int_{-1}^{1} [(1-v)\overline{V} + vu] d\overline{y} \quad (20)$$

$$v(1) = v(-1) \quad (21)$$

Here, equations (17) and (18) correspond to the no-slip boundary conditions on the walls i.e. at $\overline{y} = -1$ and $\overline{y} = 1$. Equation (19) refers to the amount of granular solid fed into the vertical channel. Equation (20) is the flow rate of the mixture and equation (21) is the symmetric boundary condition.

NUMERICAL RESULTS

The system of equations $(13)_{1,2,3}$ subjected to the boundary conditions (17) through (21) are solved numerically using a collocation code COLSYS [cf. Ascher et al. (1981)]. Collocation is implemented by COLSYS using B-spline basis functions. COLSYS also features an adaptive mesh-selection procedure based on error estimates. The mesh points are repositioned to roughly equidistribute the error, which is estimated using mesh halving and checked against user prescribed tolerances. The integral condition was implemented using a secant shooting method to refine the initial guesses. Here, we carry out a parametric study of the equations and delineate how the various non-dimensional parameters affect the structure of the solution.

The manner in which the volume fraction profiles change with B are shown in Figure 2. Notice, that as B increases the volume fraction increases toward the center of the channel and decreases towards the walls of the channel. Figure 3 shows the effect of R_2 on solid velocity profile. Here, the solid velocity increases as R_2 is increased. The manner in which R_2 effects the fluid velocity profile is shown in Figure 4. Notice, that the fluid velocity decreases as R_2 is increased.

ACKNOWLEDGMENT

The authors wish to thank the Pittsburgh Energy Technology Center, US Department of Energy for its support.

LITERATURE CITED

1. Ascher, U., Christianson, J. and Russel, R. D., " Collocation Software for Boundary Value ODE's," in *ACM Transactions on Mathematical Software*, 7/2, p. 209 (1981).
2. Bowen, R. M., "Theory of Mixtures in Continuum Physics" Edited by A. C. Eringen, Vol. 3 (1976).
3. Goodman, M. A., and Cowin, S. C., "Two Problems in the Gravity Flow of Granular Materials," in *J. Fluid Mech.*, 45, p. 325 (1971).
4. Goodman, M. A., and Cowin, S. C., "A Continuum Theory for Granular Materials," in *Arch. Rat. Mech. Anal.*, 44, p. 249 (1972).
5. Green, A. E., and Naghdi, P. M., "On Basic Equations for Mixtures," in *Q. J. Mech. Appl. Math.*, 22, p. 427 (1969).
6. Gudhe, R, Yalamanchili, R. C., and Massoudi, M.," Flow of Granular Materials Down a Vertical Pipe," in *Int. J. Non-Linear Mech.*, 29, pp. 1 (1994).
7. Johnson, G., Massoudi, M., and Rajagopal, K. R.," Flow of a Fluid-Solid Mixture Between Flat Plates," in *Chemical Engineering Science*, 46, p. 1713 (1991a).
8. Johnson, G., Massoudi, M., and Rajagopal, K. R.," Flow of a Fluid Infused with Solid Particles Through a Pipe," in *Chemical Engineering Science*, 29, p. 649 (1991b).
9. Massoudi, M., " Application of Mixture Theory to Fluidized Beds," Ph.D. Thesis, University of Pittsburgh (1986).
10. Rajagopal, K. R., and Massoudi, M.," A Method for Measuring Material Moduli of Granular Materials: Flow in an Orthogonal Rheometer," DOE/PETC/TR-90/3 (1990).

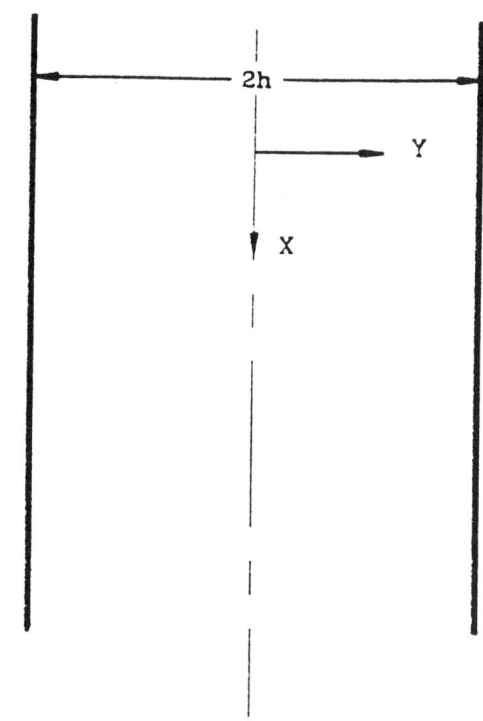

Figure 1. Flow Between Two Infinite Plates

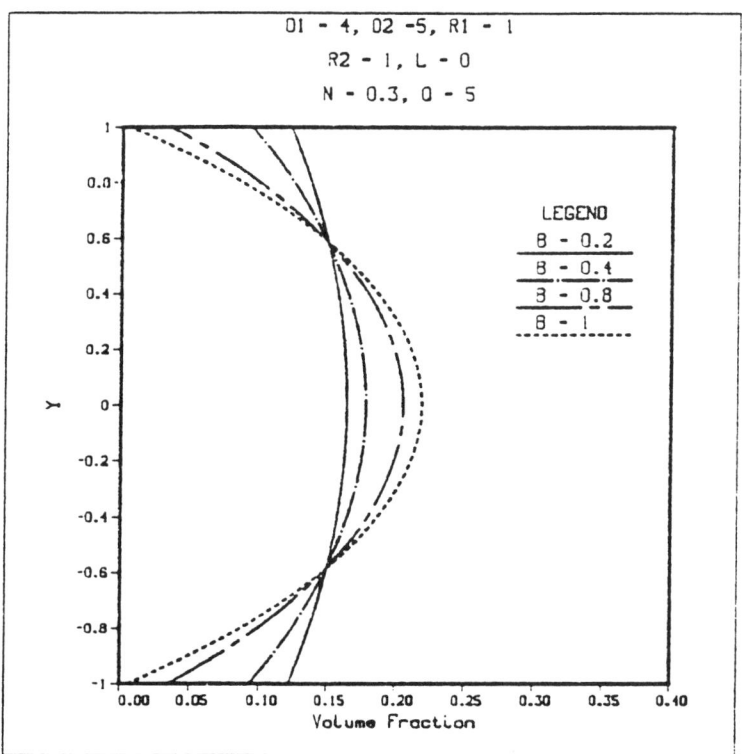

Figure 2. Effect of B on the Volume Fraction Profile

118 Developments in Fluidization and Fluid-Particle Systems

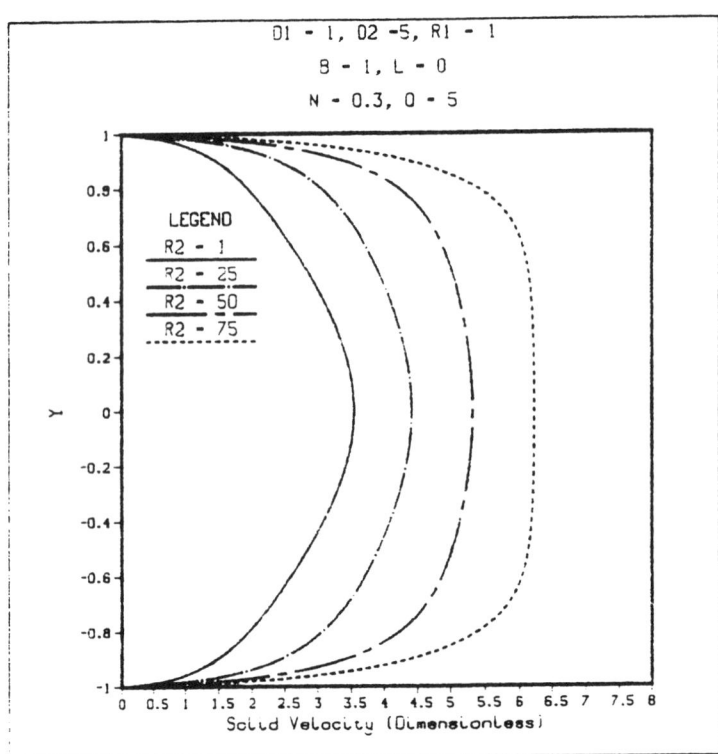

Figure 3. Effect of R_2 on the Solid Velocity Profile

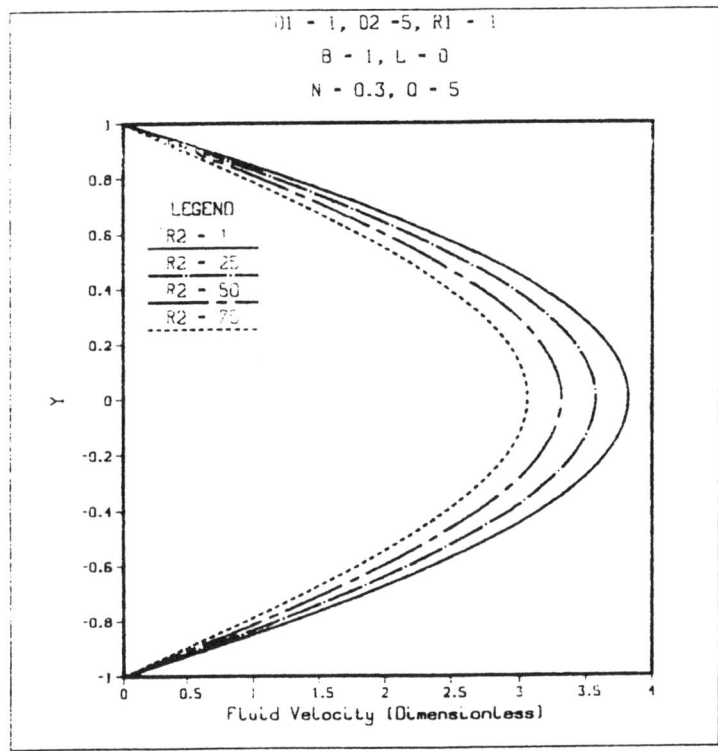

Figure 4. Effect of R_2 on the Fluid Velocity Profile

Microreactor Simulating Reaction Scene in Turbulent Fluid Bed of Group A Powder: 1. Axial Gas Dispersion

G. Gregory Benge and Arthur M. Squires
Virginia Polytechnic Institute and State University, Blacksburg, VA 24061

A horizontal duct of rectangular cross-section (height = 12.7 mm) containing powder in the coherent-expanded vibrated-bed state [1] provides a microreactor in which axial gas dispersion can be treated by a simple axial dispersion model [2,3]. We report herein axial dispersion coefficients obtained by means of pulsed-tracer experiments.

We have reported [1] how a shallow bed of particles resting on a horizontal plate that is vibrated sinusoidally in the vertical direction can display one of several "states": (1) A "vibrated bed" comprising a very few particles will present a "Newtonian state," in which particles bounce on the vibrating plate as individuals. There are two such states. (2) At a larger inventory of particles, the "coherent-expanded" state develops, wherein particles move in concert while undergoing vigorous vertical mixing and achieving order-of-magnitude increase in height of travel of particles relative to a compacted bed depth (i.e., before vibration commences). (3) At still larger particle inventory, the "coherent-condensed" state appears. In this state, the bed expands hardly at all (except, for fine particles, in a surface layer). Mixing is sluggish (except in a surface layer), and individual particles tend to remain for some time in a fixed relationship one to another.

G. G. Benge is now with Eastman Chemical Company, Kingsport, Tennessee.

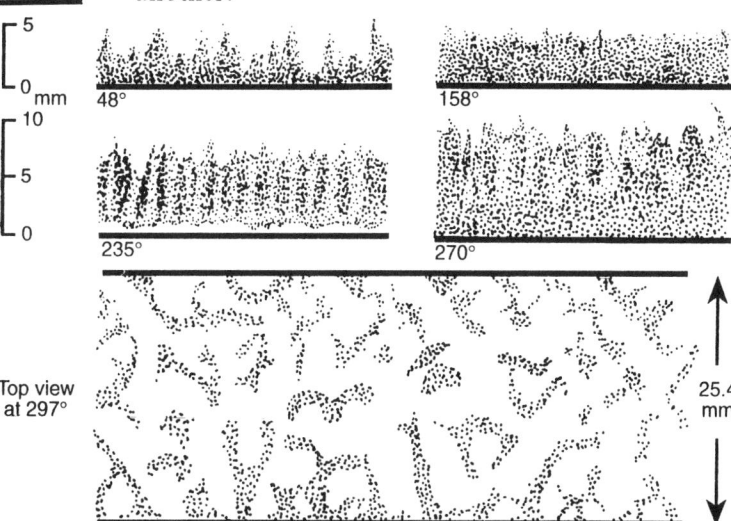

Figure 1. **Coherent-expanded state in a vibrated bed of Group A powder.**

The upper four sketches are side views. The lower sketch is a top view. The drawings are tracings from photographs [1] of a vibrated layer of FCC powder (0.96 mm in compacted state) at phase angles noted alongside the drawings. Conditions: 25 Hertz; amplitude, 1.99 mm; maximum acceleration, five times gravity. If maximum acceleration is about six to seven times gravity, the layer expands to more than 13 mm at ~270°.

MICROREACTOR

The coherent-expanded state of a Group A powder offers the opportunity we are pursuing. **Figure 1** is a series of drawings illustrating behavior of a shallow layer of fluid cracking catalyst (FCC) powder resting on the floor of a chamber vibrated vertically at 25 Hertz and maximum displacement of 1.99 mm. In the experiment, maximum floor acceleration is five times gravity. The layer contains a weighed quantity of weighed quantity of FCC powder calculated as sufficient to form a 960 μm layer in a highly compacted condition. At phase angles between ~60° and ~150°, the layer occupies a height of ~4 mm (i.e., expanded about four times its compacted depth). At ~150°, the layer begins to lift, gas penetrating the layer to create a gap between layer and plate. Between ~230° and ~40°, gas in the gap rushes upward through the powder, which expands to as much as 10 mm while reorganizing itself to form bands about a millimeter in width, at spacings of a few millimeters. This repeats 25 times in each second; powder mixing is intense in the vertical direction, while relatively small in the horizontal direction. Absolute gas velocities in the vertical direction — readily estimated from phase-shift photographs [1] — are relatively large, approaching 0.3 m/s; and so vertical gas mixing is also intense.

Our concept is to cause gaseous reactants to flow horizontally through a vibrating duct containing a coherent-expanded layer of powder. Vibration parameters would be chosen such that powder achieves full duct height once in each cycle, between ~270° and ~300° phase angle. We have reported experiments with atmospheric air showing negligible lateral displacement of a fluid cracking catalyst (FCC) powder (i.e., from duct inlet toward outlet) at horizontal air superficial velocities up to ~2.0 cm/s [4].

AXIAL DISPERSION MODEL

Simple visual observation (best done under strobe lighting at a frequency slightly less than the vibrational frequency) makes it evident that the microreactor duct provides intimate gas-solid contacting and minimal gas bypassing; also, that axial gas dispersion in the duct may be treated appropriately by a simple axial dispersion model [2, 3] having axial Peclet number, Pe_{ax}, as parameter:

$$Pe_{ax} = \frac{UL}{D_{ax}} \quad [1]$$

where L = length of duct (cm); U = horizontal superficial gas velocity (cm/s); and D_{ax} (cm²/s) is an axial dispersion coefficient (combining effects of both molecular diffusion and fluid dispersion). The Pe_{ax} of Equation 1 is a ratio of two times: time for dispersional relaxation over reactor length L (on the order of L^2/D_{ax}) divided by fluid residence time (L/U). This Pe_{ax} is sometimes called a "reactor Peclet number" [5]. It should not be confused with the "particle" or "fluid Peclet number" (in which particle diameter replaces L in Equation 1), often used in treating gas dispersion in packed beds [3, 6].

In this paper and a paper to follow [7], we report D_{ax}-values in the proposed microreactor, obtained by means of pulsed-tracer experiments employing helium, argon, and sulfur hexafluoride in nitrogen carrier. Here, we give values for a 20-cm-long duct containing glass beads of Group A character; in the paper to follow, values for an FCC powder as well as glass-bead data for shorter ducts.

The paper to follow [7] also discusses the microreactor's salient features (including efficiency of gas-solid contacting), and deals with the microreactor's usefulness in fluid-bed reaction engineering.

EXPERIMENTAL ARRANGEMENTS

The vibration system is generally similar to an earlier setup [8], which, however, employed a circular vibration table and a cylindrical steel support structure resting upon a large concrete block. The present system employs a rectangular vibration table and box-like steel support. The assembly rests upon two layers of 63.5-mm thick foam rubber padding on the floor of the work space of a fume hood (a location permitting the assembly's future use for a microreactor processing gases or vapors at elevated temperatures). Leaf springs that support the vibration table are tuned to a resonance frequency close to frequencies used in this work (~23 to ~27 Hertz). Especially important is maintenance of strict verticality in the "drive shaft" (~75-mm length of piano wire, 2.38 mm in diameter) linking the vibration table to an electromagnetic vibrator. **Figure 2** is an isometric drawing illustrating a third-generation (variable-length) microreactor model used in the present cold-flow experiments. The microreactor duct is horizontal, 25.4 mm wide, 12.7 mm high, and 20 cm long. A sliding-plug inlet-gas distributor permits the active length of duct to be varied. (In two earlier designs, duct length was fixed at 20 cm, and gas entered the duct through a porous grid in the ceiling at one end.) **Figure 3** is a schematic cross-section illustrating assembly of the microreactor test section and its mounting on the vibration table. Notice that the duct's ceiling is glass, as well as sides and floor.

Two powders have been used in the work:

(1) Soda-lime glass beads: spherical, nonporous; 53 to 74 μm (-200+270 U.S. Standard mesh); average bulk density, 1.55 g/cm^3; solid density, 2.50 g/cm^3.

(2) CBZ-1 FCC catalyst: manufactured by Davison; calcined at ~575° before use; weight fractions in various size ranges: 0 @ <20 μm, 0.07 @ 20-40 μm, 0.40 @ 40-80 μm, 0.36 @ 80-105 μm, 0.16 @ 105-149 μm, 0.01 @ >140 μm; pore volume, 0.58 cm^3/g; average bulk density, 0.74 g/cm^3, measured by compaction [*1*]; solid density, ~2.57 g/cm^3; surface area, 214 m^2/g.

In an experiment, a weighed quantity of powder is introduced into the test duct, this quantity having been calculated to be sufficient to provide a uniform layer one mm deep on the floor of the duct if the power were present in its fully compacted state [*1*]. Under the vibration conditions of this

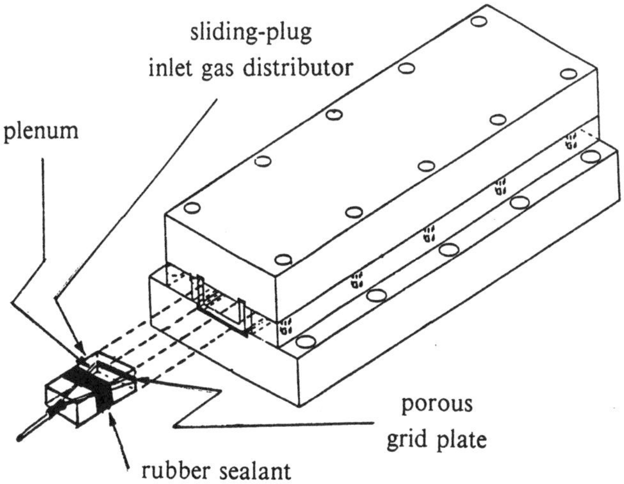

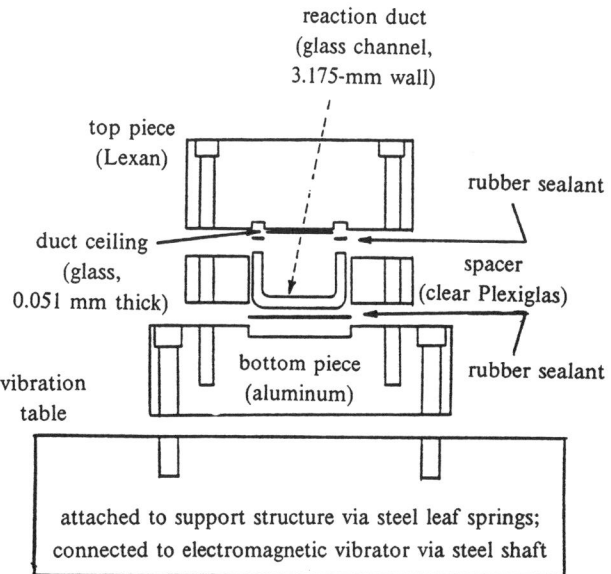

Figure 2. Isometric drawing of third-generation vibrated-bed microreactor cold-flow model.

A gas-inlet piece (with plenum zone) fits snugly inside a rectangular duct, and can be positioned at various distances from the outlet, to provide ducts of different lengths.

Figure 3. Schematic diagram illustrating assembly of vibrated-bed microreactor cold-flow model.

work (generally ~27 Hz and maximum acceleration ≃ six to seven times gravity), this layer should enter the coherent-expanded state (illustrated in Figure 1); in practice, visual observation provides a check upon the presence of this state in each experiment.

In a tracer experiment, a six-port automatic valve injects an imperfect pulse into the test duct. Molecular diffusivities (D_{mol}) for our three tracer gases in nitrogen (298°K, atmospheric pressure, and 712 mm Hg, the average pressure of our experiments) are [9]:

	@ 760 mm Hg	@ 712 mm Hg
SF_6	0.0974 cm²/s	0.104 cm²/s
argon	0.194	0.207
helium	0.690	0.737

A thermal conductivity detector (TCD) detects helium or argon; an election capture detector (ECD), SF_6. Since responses of both TCD and ECD are linear in ECD are linear in tracer concentration, tracer response ($R(t)$, where t is time) may be used as a surrogate for concentration in data analysis (except in a mass balance).

DATA ANALYSIS

Our tracer impulse undergoes dispersion before it enters the microreactor duct; accordingly, the impulse is not ideal. To obtain effective diffusivities and Peclet numbers for the duct, we employ the "imperfect impulse method" of Bischoff and Levenspiel [10], which is useful "when something at least vaguely resembling an impulse or step change can be applied at the system inlet," provided, of course, application of the method does not rely upon a small number obtained as a difference of two large numbers [11]. Figure 4 plots typical inlet and outlet responses (after application of noise reduction techniques to be discussed below).

Possessing both inlet and outlet tracer responses, we may invoke the axial dispersion model to determine Pe_{ax}. There are several ways to perform the determination [12]: the most common are (i) the classic method of moments; (ii) the weighted moments method; (iii) the transfer function method; (iv) least-squares curve fitting in the frequency domain via frequency response analysis;

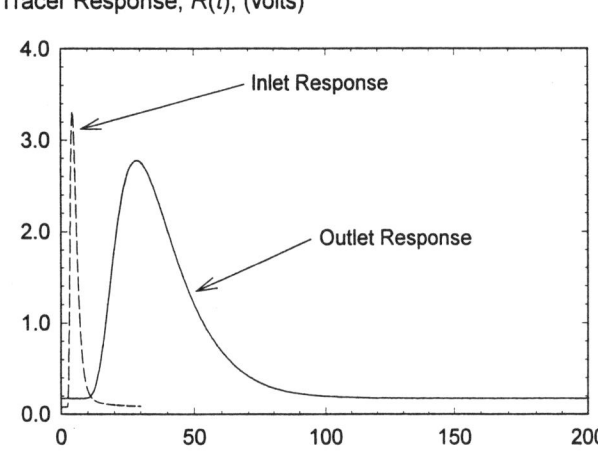

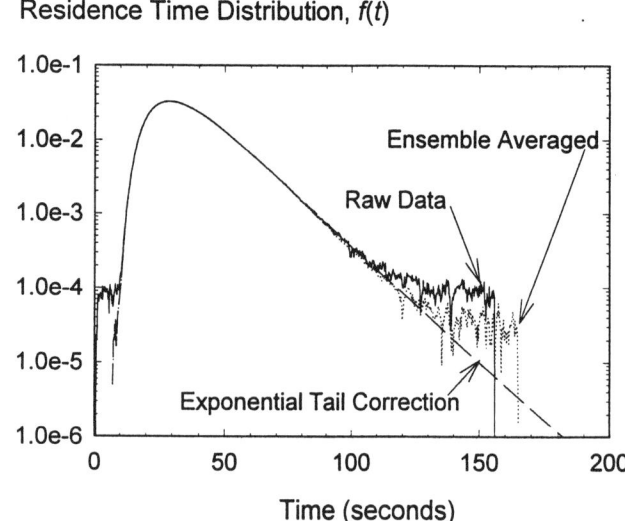

Figure 4. Examples of inlet and outlet responses, $R(t)$, ensemble-averaged from results of five or six runs. Inlet response reflects a tracer injection eight times smaller than that for outlet response.

Figure 5. Example of residence time distribution, $f(t)$, illustrating treatment of data: ensemble averaging and exponential tail correction.

and (v) least-squares curve fitting in the time domain (with either convolution or deconvolution techniques) [*11, 13–20*]. Although classic method (i) is widely used because of its simplicity, it is often criticized because it weights heavily the noisy data in the tail portion of a tracer response.

We use a modified method of moments technique, focusing effort upon minimizing error in the tail. Consider **Figure 5**, which provides typical plots comparing data before and after application of several techniques for dealing with noise. In Figure 5, the residence time distribution function $f(t)$ is calculated from tracer response, thus:

$$f(t) = \frac{R(t)}{\int_0^\infty R(t)dt} \qquad [2]$$

First, consider the raw data in Figure 5. Both run design and equipment design are tailored to contribute to accuracy. Oversampling tracer response by a factor of ten, followed by boxcar averaging, reduces noise level by roughly the square root of ten [*21, 22*]. In the data acquisition system, resolution of a 12-bit analog-to-digital converter is ±0.61 mv for signals between 0 and 5 v. Detector signal amplification closely matches input range of the converter. Experiments are designed to yield ~9 to 10,000 raw data points, with data collected to ~5 mean residence times to ensure sufficient measurements of the tail; the design dictates sampling rate, which varies from ~15 to 450 Hz. Tracer injection volume is varied so that peak output response is well matched to converter input range; for example, the exit response in Figure 4 reflects injection of eight times as much tracer as that used for the inlet response. A tracer experiment is repeated 5 or 6 times to further reduce noise, and data are ensemble-averaged [*22*] to yield an "average tracer response," $R(t)$. Averaging and noise reduction techniques were tested and found to affect only noise and reproducibility, and not absolute results per se. In tests for SF_6 at tracer concentrations of 1.1 and 11.6 ppm and at pulse volumes of 25, 50, 100, 500, and 1000 μl, no evidence of nonlinearity in the results could be seen.

We remove a background signal by subtracting values along a line connecting the lowest points before and after the peak [*21*]. We apply an exponential tail correction to remove the inherently noisy data of the tail in Figure 5. Although authors often refer to a correction of this nature [*23–25*], we found no "standard method" for practical application that we deemed appropriate for our work [*12*]. Our approach is iterative, relying primarily upon visual inspection of each successive trial; we repeat the correction until we are satisfied with the fit. From a logarithmic plot of either $R(t)$ or $f(t)$, we estimate start and end points of a linear region (the latter at a time just before the onset of noise); a linear regression between the two points provides the exponential tail correction (the dashed line in Figure 5).

We integrate $R(t)$ using an algorithm based on n-applications of Simpson's rule [*26*], and obtain a tracer mass balance:

$$m_i = Qk \int_0^\infty R(t)dt \qquad [3]$$

where m_i = quantity of tracer injected in the pulse; Q = flow rate; and k = detector calibration factor, converting detector response into tracer concentration. The mass balance provides a test (the zeroth-moment test) whether we have detected all of the tracer in gas leaving the duct.

By the aforementioned integration algorithm, we obtain the first and second moments, i.e., mean residence time $\bar{t}$ and variance σ^2, respectively:

$$\bar{t} = \int_0^\infty t f(t)dt \qquad [4]$$

$$\sigma^2 = \int_0^\infty (t - \bar{t})^2 f(t)dt \qquad [5]$$

Knowledge of flow rate Q and duct volume V permits us to calculate a second value of the mean residence time, τ, a value independent of tracer data:

$$\tau = V/Q \quad [6]$$

A good comparison of $\bar{t}$ with τ gives reassurance that the duct displays no significant dead volume, i.e., volume not traversed by the tracer.

Underlying the Bischoff-Levenspiel imperfect impulse method is the additivity property of the moments [2]. Values for duct residence time and variance (identified by the subscript "d") are obtained by subtracting values for the inlet response from those for the outlet response:

$$\bar{t}_d = \bar{t}_{outlet} - \bar{t}_{inlet} \quad [7]$$

$$\sigma_d^2 = \sigma_{outlet}^2 - \sigma_{inlet}^2 \quad [8]$$

Defining dimensionless time, $\theta = t/\bar{t}$, we have a dimensionless density function, $f(\theta) = \bar{t} f(t)$, and dimensionless variance, $\sigma_\theta^2 = \sigma^2/\bar{t}^2$. A dimensionless variance for the microreactor duct is

$$\sigma_{\theta,d}^2 = \sigma_d^2/\bar{t}_d^2 \quad [8]$$

For our "simple" physical situation (no dead zones; no gas lagging behind or running ahead of a plug flow), we may calculate an axial Peclet number from an equation derived for the axial dispersion model [2, 3]:

$$\sigma_{\theta,d}^2 = (2/Pe_{ax}) \times$$
$$[1 - (1/Pe_{ax})\{1 - \exp(-Pe_{ax})\}] \quad [9]$$

Finally, Equation 1 permits calculation of D_{ax} from a measured value of Pe_{ax}.

RESULTS FOR GLASS BEADS [DUCT LENGTH = 20 CM]

In runs with He and Ar, relative errors in comparisons of $\bar{t}$ with τ fall in general below ~5%; errors in SF$_6$ runs, below ~8%. Mass balances for the runs [12] are not as good as residence time comparisons; but, since mass balances are seldom reported in tracer studies, we have little basis for judging ours. Our SF$_6$ balances are best, usually within ~5%. A tendency in helium and argon balances seems to reflect a bias toward ~15% understatement of inputs. Exponential tail corrections afforded large reductions in standard deviations in averages of data from five or six runs. For example, in runs with He, the reduction ranges from ~2-fold at low gas velocity to ~20-fold at higher [12].

In **Figure 6**, the upper three lines give our axial dispersion coefficients versus duct velocity. Data for each of the three tracers appear to define a straight line, and each line appears to have a common slope. As the lower two lines in Figure 6 demonstrate, we discovered a similar situation in Suzuki and Smith's data for axial dispersion in a fixed bed of small glass beads [6]. Having studied a far wider range of gas velocity than we are able to achieve in our system, these authors presented results in logarithmic plots of D_{ax}/D_{mol} versus particle Peclet number. Our interest is in their data at low velocities, comparable to ours, and the arithmetic plots of Figure 6 are more appropriate here.

In the two lower lines of Figure 6, each intercept at zero velocity reflects simply that mixing which arises from molecular diffusion. Each intercept $= \eta D_{mol}$, where η is a constant. Simple geometric consideration [27] shows that η is void fraction divided by tortuosity; i.e., $\eta \simeq 0.43/\sqrt{2} \simeq 0.3$ for a fixed bed.

In **Figure 7**, the lower line illustrates how Suzuki and Smith's low-velocity data can be collapsed to a common straight line passing through zero, by subtracting ηD_{mol} from measured values of D_{ax}, where $\eta = 0.3$. The lower line in Figure 7 implies an expression for D_{ax} in a fixed bed at low fluid velocity:

$$D_{ax} = \eta D_{mol} + WU \quad [10]$$

where W is a proportionality constant. In Equation 14, the first term reflects molecular diffusion; the second term, a "Wilhelm dispersion mechanism" arising from interaction of the flow with the particles of the fixed bed [3, 28].

As the upper line in Figure 7 illustrates, we discovered that our three sets of data, too, could be collapsed to a common straight line by subtracting from each set a quantity proportional to the relevant molecular diffusion coefficient. The upper line implies:

$$D_{ax} = D_{vib} + \eta D_{mol} + WU \quad [11]$$

$$D_{ax} = 0.20 + 1.1 D_{mol} + 0.257 U \quad [11A]$$

The quantity D_{vib} reflects a dispersion associated with motion of gas dragged upward and downward by particles moving in trajectories that characterize the coherent-expanded vibrated-bed state (see Figure 1).

The surprise is the empirical value, $\eta = 1.1$, that we found to be required for collapsing our three sets of data onto one line. Although, by analogy with the fixed-bed case, one might expect $\eta \cong 0.95/1.05 \cong 0.9$, we can readily advance an argument that this analogy gives an underestimate. In Figure 1, notice how particles, at phase angles between ~200° and 300°, become sorted into vertical strands. Although these are highly dilute, nevertheless, concentration gradients may arise between strand interiors and adjacent spaces that are essentially vacant. The time scale for diffusional relaxation of such a gradient is p^2/D_{mol}, where p = path length for the diffusion. At p = 0.5 mm and $D_{mol} = 0.737$ cm²/s (the value for helium), diffusional relaxation time is 0.0034 seconds, about 8.5% of the 0.04-s period of the vibration at 25 Hertz. Therefore, diffusional relaxation occurs while phase angle passes through $(0.085)(360) \simeq$

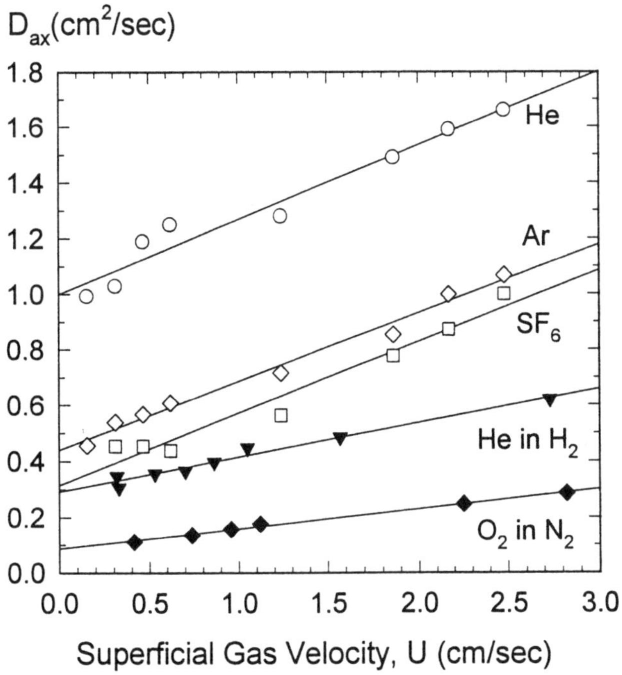

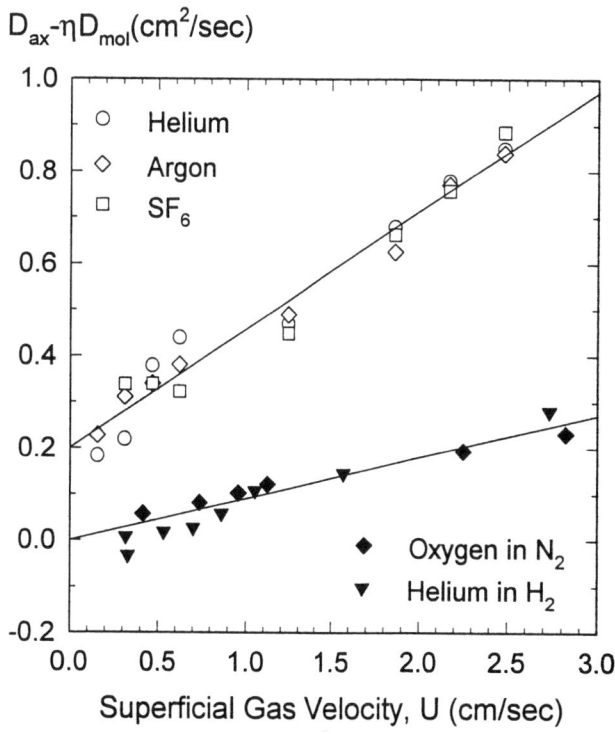

Figure 6. **Axial dispersion in microreactor and fixed bed.** Open symbols signify our data. Closed symbols signify Suzuki and Smith's data [6] for fixed beds of ~0.5-mm glass beads.

Figure 7. **Axial dispersion minus effect of molecular diffusion.** For upper line, $\eta = 1.1$; for lower, $\eta = 0.3$.

30°, considerably less than the life of a strand. The calculation suggests how diffusional relaxation between strand and empty space can make a more significant contribution to the overall gas dispersion in an experiment with helium than in one with SF_6. Furthermore, as seen in Figure 1, powder in the strands falls back to form a roughly 4-mm layer that endures through more than 100° of phase angle. In an experiment with helium, diffusional relaxation occurs in each cycle between the empty duct space above the layer and an upper region of the layer, ~1 mm thick; moreover, there is significant relaxation between a similar lower region of the layer and the empty space of the gap that forms below the layer after ~158°. We note that gas forming the gap does not sweep downward through the layer in a uniform plug flow; rather, the layer becomes dimpled, with gas rushing toward the floor at each dimple [1].

DISCUSSION

Although all measurements underlying Equation 11A were made in a 20-cm-long duct, a D_{ax} value predicted by the equation reflects the microstructure of the coherent-expanded vibrated-bed state in our glass beads. It is reasonable to suppose that this D_{ax} value holds good in ducts of widely varying length. Data for shorter ducts, to be presented in the paper to follow [7], demonstrate that this is indeed the case.

The paper to follow also demonstrates how a user of the microreactor may vary Pe_{ax} in a series of experiments, simply by varying duct length and superficial gas velocity in proportion, with substantially no effect upon gas-solid contacting efficiency.

ACKNOWLEDGMENTS

National Science Foundation joint industry-university research grant CBT-8620244 supported the work. We are grateful to our collaborators, Frederick J. Krambeck and Amos A. Avidan of Mobil Research & Development Corporation, for their interest, comments, and suggestions. Benku Thomas designed and built the vibrated-bed support assembly and made intellectual contributions. Riley T. Chan assisted in both design and construction of equipment — notably, for automation and data acquisition. Ray Dessy was generous with help and advice on automation and data acquisition.

NOTATION

D_{ax}	effective axial diffusivity (cm²/s) (combining effects of molecular diffusion and motion of particles induced by vibration)
D_{mol}	molecular diffusivity (cm²/s)
D_{vib}	dispersion from vibration, Equation (11)
$f(t)$	residence time distribution function, Equation (2)
L	length of microreactor duct (cm)
Pe_{ax}	axial ("reactor") Peclet Number, UL/D_{ax}
Q	duct volume (cm³)
$R(t)$	tracer response
t	time (s)
$\bar{t}$	mean residence time from tracer response, Equation (4)
t_d	mean duct residence time, Equation (7)
t_{inlet}	mean residence time from inlet tracer response
t_{outlet}	mean residence time from outlet tracer response
U	horizontal superficial gas velocity (cm/s)
V	volumetric flow rate (cm³/s)
W	proportionality constant (cm), Equations (10) and (11)
WHSV	weight hourly space velocity [weight flow of reaction gas per hr]/[weight of catalyst]
η	proportionality constant (dimensionless), Equations (10) and (11)
σ^2	duct residence time variance, Equation (5) (s²)
σ_d^2	residence time variance, Equation (8) (s²)
σ_{inlet}^2	residence time inlet response variance (s²)
σ_{outlet}^2	residence time outlet response variance (s²)
$\sigma_{\theta,d}^2$	dimensionless duct variance, Equation (8)

τ mean residence time from V/Q (s)

θ dimensionless time, $t/\bar{t}$

LITERATURE CITED

1. Thomas, B., M.O. Mason, Y.A. Liu, and A.M. Squires, *Powder Technol.*, **57**(1989):267-280.
2. Levenspiel, O., *Chemical Reaction Engineering*, Wiley, New York, 2nd edition, 1972.
3. Wen, C.Y., and L.T. Fan, *Models for Flow Systems and Chemical Reactors*, Marcel Dekker, New York, 1975.
4. Thomas, B., and A.M. Squires, "Vibrated-Bed Microreactors Simulating Catalytic Fluid Beds: A Feasibility Study" in *Fluidization VI*, J.R. Grace, L.W. Shemilt, and M. Bergougnou (Eds.), Engineering Foundation, New York, 1989, pp. 375-382.
5. Fogler, H.S., *Elements of Chemical Reaction Engineering*, Prentice-Hall, Englewood Cliffs, New Jersey, 1986.
6. Suzuki, M., and J.M. Smith, *Chem. Eng. J.*, **3**(1972):256-264.
7. Squires, A.M., and G.G. Benge, THIS VOLUME.
8. Thomas, B., Y.A. Liu, R.T. Chan, and A.M. Squires, *Powder Technol.*, **52**(1987):77-92.
9. Bird, R.B., W.E. Stewart, and E.N. Lightfoot, *Transport Phenomena*, Wiley, New York, 1960.
10. Bischoff, K.B., and O. Levenspiel, *Chem. Eng. Sci.*, **17**(1962):245-255.
11. Nauman, E.B., and B.A. Buffham, *Mixing in Continuous Flow Systems*, John Wiley & Sons, New York, 1983.
12. Benge, G.G., "Cold model of vibrated bed microreactor ...", Ph.D. thesis, Virginia Polytechnic Institute & State University, Blacksburg, 1992.
13. Ostergaard, K., and M.L. Michelsen, *Can. J. Chem. Eng.*, **47**(1969):107-112.
14. Michelsen, M.L., and K. Ostergaard, *Chem. Eng. Sci.*, **25**(1970):583-592.
15. Michelsen, M.L., *Chem. Eng. J.*, **4**(1972):171-179.
16. Michell, R.W., and I.A. Furzer, *Chem. Eng. J.*, **4**(1972):53-63.
17. Felder, R.M., R.E. Harrison, and R.W. Rousseau, *Chem. Eng. Commun.*, **1**(1974):187-189.
18. Fahim, M.A., and N. Wakao, *Chem. Eng. J.*, **25**(1982):1-8.
19. Mills, P.L., and M.P. Dudukovic, *Computers Chem. Engng.*, **13**(1989):881-898.
20. Buffham, B.A., and G. Mason, *Chem. Eng. Sci.*, **48**(1993):3879-3887.
21. Gates, S.C., and J. Becker, *Laboratory Automation Using the IBM PC*, Prentice Hall, Englewood Cliffs, New Jersey, 1989.
22. Dessy, R.E., *Laboratory Automation*, American Chemical Society, Washington, D.C., revised edition, 1986.
23. Dudukovic, M.P., and R.M. Felder, "Mixing in Chemical Reactions — V — Micromixing and the Segregated Flow Model" in AIChE Modular Instruction, Series E: Kinetics, Volume 4: Reactor Stability, Sensitivity and Mixing Effects, AIChE, New York, 1983, pp. 62-70.
24. Waldram, S.P., "Non-Ideal Flow in Chemical Reactors" in *Comprehensive Chemical Kinetics*, G.H. Bamford et al. (Eds.), Elsevier, Amsterdam, 1985, vol. 23, pp. 223-281.
25. Dudukovic, M.P., "Tracer Methods in Chemical Reactors. Techniques and Applications" in *Chemical Reactor Design and Technology (NATO ASI Ser., Ser. E.*, vol. 110), H.I. de Lasa (Ed.), Martinus Nijhoff Publishers, Dordrecht, 1986, pp. 107-189.
26. Carnahan, B., H.A. Luther, and J.O. Wilkes, *Applied Numerical Methods*, Wiley, New York, 1969.
27. Edwards, M.F., and J.F. Richardson, *Chem. Eng. Sci.*, **23**(1968):109-123.
28. Wilhelm, R.H., *Pure Appl. Chem.*, **5**(1962):403-421.

Microreactor Simulating Reaction Scene in Turbulent Fluid Bed of Group A Powder: 2. Usefulness in Fluid-Bed Reaction Engineering

Arthur M. Squires and G. Gregory Benge
Virginia Polytechnic Institute and State University, Blacksburg, VA 24061

A vibrated-bed microreactor exploiting the coherent-expanded state [1] is capable of simulating (independently) two important features of the reaction scene in a turbulent fluid bed of a Group A powder: axial gas dispersion and top-to-bottom solid circulation. Microreactor data can influence a decision whether to attempt development of a fluid bed or a fixed bed for conducting a reaction.

For a newly discovered heterogeneously catalyzed reaction, how sensitive are reaction outcomes (conversion, selectivities, etc.) to axial gas dispersion? Engineers planning a reactor development may well ask themselves this question.

If the new reaction is highly exo- or endothermic, or requires frequent replacement of catalyst, an institution's managers may well authorize an initial stage of a research and development (R&D) whose ultimate goal is design of a turbulent fluid bed reactor employing a Group A catalytic powder [2]. In general, this stage comprises scouting studies whose major aim is to show whether a fluid bed affords an appropriate environment for the reaction. From the work, engineers learn the approximate weight hourly space velocity (WHSV) necessary to achieve a desired level of conversion. Usually, a laboratory's budget constrains quantities of reactants and catalysts available for the work: laboratory fluid beds are commonly ~12 to 24 mm in diameter and at most a few meters tall; and gas velocity is commonly well below a level required to achieve turbulent fluidization of a Group A powder. An institution routinely engaged in chemical reaction engineering R&D can easily spend each year a dollar sum in six or seven figures on experiments in laboratory beds that operate at unrealistic fluidization parameters.

In these beds, whether slugging or turbulent, axial gas dispersion is relatively small. Since axial dispersion, other things being equal, increases with bed diameter [3], the question of a reaction's sensitivity to axial gas mixing arises if a fluid-bed R&D is to proceed beyond the laboratory-scale stage. In a large fluid bed, engineers can call upon several means for reducing axial dispersion: e.g., by raising gas velocity, increasing the level of fines in catalyst powder, increasing height-to-diameter of bed, or installing horizontal baffles (at some cost in temperature uniformity) [2, 4]. When engineers possess only laboratory results, however, few managements would appropriate moneys for constructing a plant whose centerpiece is to be a fluid-bed reactor.

Often, management sees great advantage in bringing a new reaction into production quickly. If a fixed-bed design is not out of the running — if

---

Virginia Polytechnic Institute & State University, Blacksburg. G.G. Benge is now with Eastman Chemical Company, Kingsport, Tennessee.

its cost is only moderately greater, and its process efficiency only moderately less — a fluid-bed R&D effort may well never proceed beyond the first, laboratory stage. Not only will the next, pilot stage be costly, but also the work delays a commercial decision. Indeed, it may not fully support such a decision; at further cost and delay, management may require data from a demonstration-scale fluid bed. A fixed bed can usually be designed with confidence from bench-scale data. Early choice of bird-in-hand fixed bed may cancel bird-in-bush fluid-bed R&D, unless the fluid bed's advantages are overwhelming.

Industry operates many fixed-bed processes that might well have employed fluid beds had quick and inexpensive means been available to ascertain a reaction's sensitivity to axial gas mixing. A recent example is the gasoline-synthesis plant at Motunui, New Zealand, employing Mobil's methanol-to-gasoline process. Commissioned in early 1986, the plant employs fixed beds designed from a pilot fixed bed producing 4 barrel per day [2]. When the plant was designed, data available at the same rate of production from a pilot fluid bed (102 mm in diameter) suggested that fluid beds would be the better choice. New Zealand, however, was unwilling to wait for the demonstration-scale fluid-bed results that subsequently became available (from a 600-mm bed) confirming the fluid bed's superiority [5, 6].

An objective of fluidization research should be to enable a large project like New Zealand's to consider the fluid bed without waiting for results from a demonstration plant.

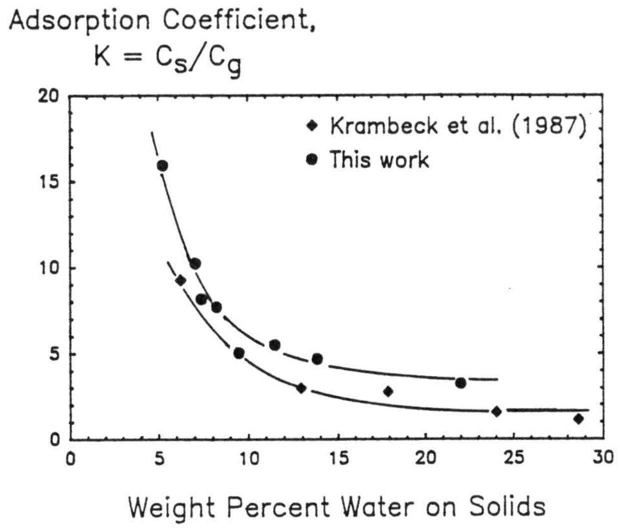

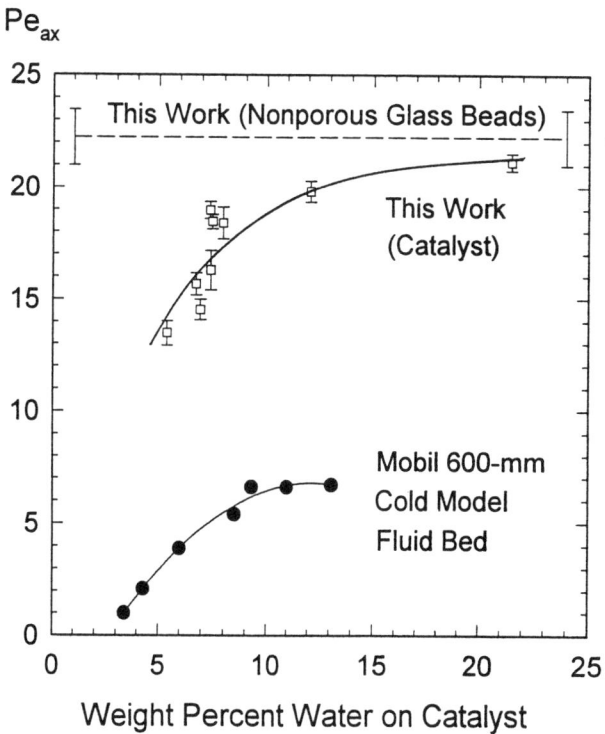

Figure 1. **Sulfur hexafluoride adsorptivity on two FCC catalysts at room temperature.** C_s and C_g are SF_6 concentrations (mass/volume) in solid and gas phases, respectively. Earlier studies in a 600-mm-diameter cold turbulent fluid bed model [9–12] used CCZ-11 catalyst; our catalyst is CBZ-1.

Figure 2. **Sulfur hexafluoride dispersions data at WHSV = 1.** Dashed line = glass beads in dry atmosphere. Curves = FCC catalyst at varying moisture level. Error bars denote data from a 20-cm-long microreactor. Closed circles denote data from 600-mm cold turbulent fluid bed model without baffles at beds heights from ∼8 to ∼13 m [9–12].

AXIAL DISPERSION MODEL

In a commercial-scale turbulent fluid bed of a Group A powder, heterogeneities are far smaller in scale than bed diameter; there is little, if any, large-scale bubbling. A literature has emerged [7–13] arguing that a simple axial dispersion model [14–15] is useful for treating gas dispersion in such a bed. The parameter of the model is axial Peclet number, Pe_{ax}:

$$Pe_{ax} = \frac{UL}{D_{ax}} \quad [1]$$

where L = height of bed (cm or m); U = superficial gas velocity (cm/s or m/s); and D_{ax} is an axial dispersion coefficient (cm²/s or m²/s).

VIBRATED-BED MICROREACTOR

The preceding paper [1] describes a vibrated-bed microreactor for which the axial dispersion model is also appropriate for treating axial gas dispersion (with L = length of microreactor duct). The paper gives D_{ax}-values for the microreactor (at L = 20 cm) charged with glass beads of a Group A character. We here report: (1) results for glass beads at L = 8.64, 16.0, and 19.2 cm; and (2) results for a fluid cracking catalyst (FCC) powder at L = 20 cm. We compare the latter findings with similar data from a fluid bed cold model at a diameter of 600 mm.

With a flat floor, the microreactor duct provides little solid mixing. The floor can be modified, however, with serrations that promote front-to-back, "racetrack" circulation of powder.

Thus, the proposed microreactor is capable of simulating two important features of the reaction scene in a turbulent fluid bed of a Group A powder: **axial gas dispersion** and **top-to-bottom solid circulation.**

RESULTS FOR FCC POWDER

Using SF$_6$ tracer and an FCC catalyst [1], we performed essentially the same experiment in our 20-cm-long microreactor as that reported for a turbulent fluid bed cold model, 600 mm in diameter [9–12]. As in the earlier work, we varied percent water on catalyst. Since water adsorbs more strongly than SF$_6$, preadsorption of water blocks adsorption sites that would otherwise be available for SF$_6$

Figure 1 compares values of the adsorption coefficient for our CBZ-1 catalyst with that for CCZ-11 catalyst used in the earlier work.

At WHSV = 1, we observed Pe_{ax} over a range of catalyst water content. The upper solid curve in **Figure 2** is a plot of the data. Each point is an average of 5 or 6 runs; error bars in Figure 2 indicate standard deviations.

The lower solid curve in Figure 2, gives data for the 600-mm cold model fluid bed. In the two curves, trends versus percent water are similar; at high catalyst water content, each curve approaches a value appropriate for a non-adsorbing powder. Notice the enormous difference in scale: microreactor D_{ax}-values are stated in cm²/s; fluid bed values, in m²/s. In the fluid bed, unlike in the microreactor, molecular diffusion has negligible effect upon axial dispersion.

In Figure 2, the horizontal dashed line denotes a microreactor Pe_{ax}-value from experiments injecting SF$_6$ tracer into glass beads in a dry atmosphere. Values for SF$_6$ tracer and catalyst at progressively higher water content approach the glass-bed value asymptotically, as they should.

For helium tracer, **Figure 3** compares Peclet numbers measured for glass beads and dry CBZ-1 catalyst. One expects glass-bead values to fall above those for catalyst, since the internal porosity of the latter, in principle, should contribute to gas dispersion. The two curves in Figure 3 fall contrary to expectation. An explanation for this surprise may be that the coherent-expanded states of the two powders differ in details of both structure and powder motion. In any event, internal porosity of catalyst appears to contribute little to dispersion under conditions of our experiments; wet CBZ-1 catalyst gives Pe_{ax} values ~3% higher than dry.

SALIENT FEATURES OF MICROREACTOR

First: An argument can be given strongly suggesting that **efficiency of gas-solid contacting is not only high but also substantially independent of horizontal gas velocity.**

One leg of the argument rests upon our study of mass transfer in aerated vibrated beds [16]. In the study, our main interest was in beds at a depth of 24 mm. Particles were "Master Beads" (nearly spherical crude alumina) and low-density glass beads, coated with naphthalene, at diameters from 125 to 841 μm. Transfer of naphthalene to nitrogen was determined by gas chromatography in absence of vibration and with vibration at intensities affording maximum accelerations of two and four times gravity. Nitrogen flows were both below and above minimum gas-fluidizing velocity in absence of vibration. At 24-mm bed depth, vibration can enhance mass transfer by as much as a factor of four.

Our interest here is in our mass-transfer data from experiments at a depth of 1 mm. These data produced a surprise: **in absence of vibration**, mass transfer at this ultra-shallow depth can be 30 to 50 times greater than in a 24-mm-deep bed. Vibration of the 1-mm layer can further increase mass transfer by as much as a factor of two, if the vibration induces the coherent-expanded vibrated-bed state.

Data for 1-mm aerated vibrated beds strongly suggest that mass transfer coefficients in the proposed microreactor are unusually large.

A second leg of the argument derives from simple observation. In the microreactor, vertical gas and particle velocities, cyclically upward and downward, are far higher than an allowable horizontal velocity. Clumps of particles form and disintegrate after a few hundredths of a second. Efficiency of gas-solid contacting for each individual particle in the microreactor is surely no less than, and may well be considerably greater than, average efficiency in a turbulent fluid bed of a Group A powder, where a clump of particles tends to remain in close company for a time interval on the order of at least a few tenths of a second, and with relatively small movement of gas through the clump's interstices.

Second: In a microreactor experiment, it is not always possible to match both WHSV and gas residence time appropriate for a commercial-scale turbulent fluid bed. This is not a disadvantage for many situations of R&D interest. **When reaction rate is a function of the available internal surface of a powder, kinetic behavior depends primarily upon WHSV.** The locus of many heterogeneously catalyzed reactions is internal surface of catalyst particles, and only WHSV has much significance, if mass transfer to external surface of a catalyst particle is not a rate-limiting factor.

Third: Heat transfer between walls, floor, and ceiling of the duct and its interior is characterized by a high heat transfer coefficient: a modest value for the coefficient would be ~300 watts/m^2-K [17].

Fourth: Before an experiment is performed for a given reaction, a prudent upper limit must be set upon horizontal superficial gas velocity, beyond which there is danger that the flowing gas might shift powder toward the exit of the microreactor

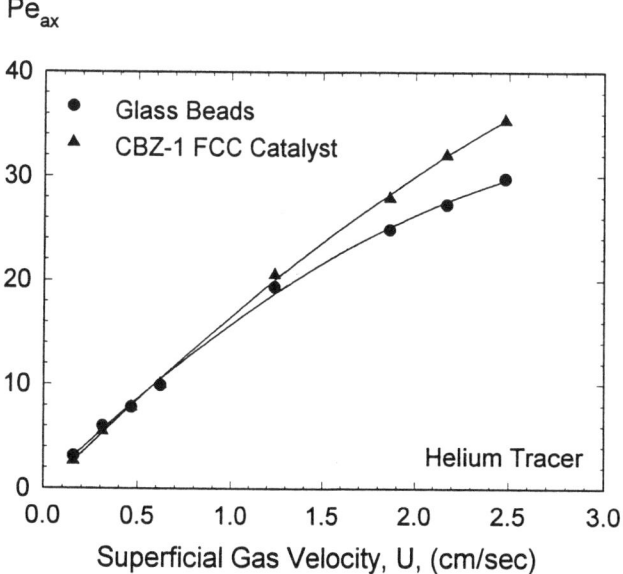

Figure 3. Axial Peclet numbers for helium tracer, in glass beads and dry FCC catalyst.

duct. Using atmospheric air and a typical, equilibrium FCC catalyst, we were unable to detect a visible shift in catalyst toward the exit until the superficial air velocity exceeded 2.0 cm/s [18].

Consider, for example, the methanol-to-olefin (MTO) reaction, conducted at 480 C, atmospheric pressure, and WHSV = 1 [19]. Scaling by viscosity ratio from 2 cm/s for air, we estimate the maximum prudent experimental velocity for this reaction to be ~1.4 cm/s. At a ZSM-5 inventory (MTO catalyst) affording a 1-mm-deep compacted layer (in absence of vibration), this maximum prudent velocity would permit an experiment at WHSV = 1 in a 44-cm-long duct. This duct affords, as we shall see below, a Pe_{ax}-value beyond ~50; i.e., the duct affords something close to plug flow.

Many fluid-bed process operate at a WHSV on the order of unity. For most of these processes, our proposed microreactor should be capable of providing data at a high Pe_{ax} value (for comparison with data at lower values). An exception, where microreactor duct length could be disappointingly restricted, might be a process working at atmospheric or sub-atmospheric pressure with a gas of unusually high viscosity, or with a catalyst of unusually low density. In a few instances such as these, the highest attainable Pe_{ax} might not permit experiments in which this parameter is varied over a sufficiently wide range.

Where a process operates at a WHSV much greater than 1, consideration must be given to employing the microreactor with a catalyst bed much shallower than 1 mm (compacted). Such consideration might well include a decision to perform additional tracer studies for the shallower bed.

At either high or low WHSV, if the microreactor is put to routine use, a laboratory can set up a simple experiment [18] disclosing the maximum prudent velocity when air is passed over a new catalyst. The maximum velocity for a new reaction gas at the desired WHSV can then be scaled by viscosity ratio.

Fifth: There is little lateral movement of solid in the microreactor (unless the working velocity greatly exceeds the maximum prudent value). The microreactor provides insignificant longitudinal solid mixing, whereas powder in a typical fluid bed circulates from top to bottom and return. The microreactor may be modified, however, to provide for front-to-back powder circulation.

EFFECT OF AXIAL GAS MIXING

The preceding paper [1] reports an empirical expression for D_{ax} in the proposed microreactor:

$$D_{ax} = 0.20 + 1.1 D_{mol} + 0.257 U \quad [2]$$

where D_{ax} and D_{mol} (molecular diffusivity) are in cm²/s; and velocity U is in cm/s.

At 1-mm compacted catalyst depth and 12.7-mm duct height, and with $S = U/L$, Equations 1 and 2 combine to give

$$Pe_{ax} = \frac{SL^2}{0.20 + 1.1 D_{mol} + 0.257 SL} \quad [3]$$

where $S = 0.1 \rho_{cat} \text{WHSV}/(1.27 \times 3{,}600 \rho_{gas})$; ρ_{cat} and ρ_{gas} are densities of catalyst (compacted) and reaction gas, respectively.

It is evident that an experimenter may explore a wide range of Pe_{ax} by conducting a reaction at constant WHSV in several microreactors of various lengths.

RESULTS FOR GLASS BEADS [DUCT LENGTH VARYING]

At WHSV = 1, and using the sliding-plug microreactor of Figure 1 of the preceding paper [1], we measured Peclet numbers for helium tracer and glass beads at three lengths of duct, to obtain the following comparison of measured axial Peclet numbers with those predicted from Equation 3:

		Axial Peclet Number	
Length	Velocity	Predicted	Measured
19.2 cm	0.6 cm/s	10.3	10.0
16.0	0.5	7.3	7.4
8.64	0.27	2.3	1.9

The good comparison provides experimental confirmation of Equation 3. The comparison demonstrates that an estimate of D_{ax} from Equation 2 is a local property, independent of duct length, solely reflecting details of the local microstructure of the coherent-expanded state.

PLANNING EXPERIMENTS

To develop an example of Equation 3's application for planning a series of microreactor experiments at several lengths of duct, again consider the methanol-to-olefin reaction. At 480°C and atmospheric pressure, average D_{mol} of methanol and reaction products is ~0.5 cm²/s. Compacted density of the ZSM-5 catalyst used in the reaction is 0.7534 gm/cm³. At WHSV = 1, S = 0.0318 s⁻¹. For several duct lengths, the following tabulation lists velocities, axial Peclet numbers, and dimensionless standard deviation in residence time distribution, $\sigma_{\theta,d}$ (i.e., standard deviation expressed as a fraction of the mean residence time of reaction gases):

Length	Velocity	Pe_{ax}	$\sigma_{\theta,d}$
40 cm	1.27 cm/s	47.2	0.20
25	0.79	20.8	0.30
15	0.48	8.2	0.46
10	0.32	3.8	0.63

The reason for the lower Pe_{ax} in a shorter duct is apparent: the tabulated $\sigma_{\theta,d}$ values illustrate how a pulse of tracer would expand to occupy a larger portion of a shorter duct.

As we have seen above, duct velocity at L = 40 cm is prudent from standpoint of danger that the horizontal flow of gas will produce horizontal displacement of powder.

For a large turbulent fluid bed of Group A powder, without baffles, an engineer may expect Pe_{ax} to be ≤ 7 [7, 9]. See lower curve in Figure 2.

If heat transfer is assumed at 300 watts/m²-K, and if most of the MTO reaction heat is liberated in ~10% of the duct, near its front end, the temperature difference between duct interior and surrounding surfaces might rise to ~10° near the front end of a uniformly heated duct.

If, for a given reaction, temperature uniformity should become a serious problem, differential heating of several portions of the duct might be provided, along with controls based upon thermocouple readings within the duct, in order to maintain the temperature profile within desired limits. Note, however, that in a series of experiments at constant WHSV in ducts of various lengths, the temperature profile, relative to length, is substantially a constant.

DESIGN FOR POWDER CIRCULATION

We observe a "racetrack" circulation of powder in a microreactor duct having a serrated floor, with serrations in one longitudinal half of the floor disposed to advance solid toward the inlet, and in the other half, toward the outlet. In the experiment, each serration is a 30°-60° right triangle, disposed with hypotenuse resting on the duct floor. The length of a serration is 0.26 mm. At 25 Hertz and a maximum acceleration ~6 times gravity, we see an apparent solid velocity of ~2 cm/s. It is evident that the rate of the circulation may be adjusted by altering the pitch of the serrations.

The role of top-to-bottom mixing of powder in fluid-bed reaction kinetics, in general, occupies the fluid-bed design engineer's mind far less than the role of axial gas mixing. This engineer, at the very least, possesses data from a laboratory bed. Whatever role top-to-bottom circulation of powder may play in the kinetics of a given reaction, these data reflect this role, at least in some degree. Top-to-bottom mixing increases with size of bed; but, so far as we are aware, this increase is responsible for no major embarrassments in fluid-bed-reactor design. Nevertheless, in principle, a few erroneous judgments have perhaps been drawn from comparisons of laboratory- and pilot-scale data. Differences in the two sets of data may have been attributed to effects of axial gas dispersion that, in fact, reflect the considerably different time-averages for travel of a typical particle between bottom and top of the smaller and larger beds. The travel brings bed particles into contact, quasi-cyclically, with both fresh feed and highly converted gas. The fundamentally unsteady-state character of fluid-bed reaction kinetics can give rise to effects that are either hurtful or helpful [2]. An understanding of effects of solid mixing upon reaction outcomes can at least increase confidence in a design, and may do more. Relatively simple means may be devised for altering the rate of top-to-bottom solid circulation. With understanding, occasions may arise where a designer will advantageously adjust this variable, upward or downward. Engineering opportunities for exploiting kinetic transients remain relatively unexplored.

DESIGN FOR HOT EXPERIMENTS

Work now in progress [20] employs a microreactor designed for temperatures to ~500°C in a study of the methanol-to-olefin reaction. The design generally resembles that shown in Figure 2 of the preceding paper [1], but with all-aluminum construction and with provision for inserting electric cartridge heaters and mounting external insulation. The design, although it has provided good data, is awkward in use; and we are studying an alternative design in which the vibrated duct would weigh far less and would be housed within a stationary furnace.

DISCUSSION

In connection with Equation 3's usefulness for planning a series of experiments with a given reaction, there are two cases to consider: (1) The reaction is newly discovered: little is known. (2) The reaction is well studied: reaction mechanisms and paths are understood, and kinetic information is available for significant paths.

For the first case, the proposed microreactor can be immediately useful as a tool providing quantitative information on the effect of a large reduction in Pe_{ax} upon outcomes of the new reaction. No prior knowledge of reaction mechanisms and kinetics is needed to obtain information of considerable worth for further R&D planning.

For the second case, the microreactor can provide results that test one's knowledge of a reaction. The test would be whether one can use this knowledge in an analysis successfully predicting the effect of a large reduction in Pe_{ax} upon a reaction outcome. A success increases one's assurance that knowledge of the reaction is sufficient for confident design of equipment and means for its control during operation.

For ideal execution of the analysis, further development work on the microreactor could be useful.

First: Equation 2, used in deriving Equation 3 for Pe_{ax}, originates from tracer experiments conducted for glass beads of Group A character [1]. In our discussion of Figure 3, we have noted the possibility that details of the structure of the coherent-expanded state may differ for various powders. Accordingly, the numerical coefficients in Equation 2 may well depend in some degree upon a powder's character. An analyst may fear that a prediction of Pe_{ax} may be flawed as an input to analysis. An improvement could result from additional tracer studies executed for the actual catalyst appropriate for a reaction under analysis.

There is another respect in which Pe_{ax} prediction could be improved. In our prediction for the MTO reaction, we used an average D_{mol} value. This could be improved upon by taking into account the variation in D_{mol} along the reaction path.

Second: If a reaction is highly exothermic or endothermic, some inequalities in temperature may arise along the microreactor duct, unless precautions are taken, by means of uneven heating, to hold temperature uniform. For microreactors of various lengths working at constant WHSV, a temperature profile is substantially a constant (relative to length), and its presence may not vitiate a conclusion concerning the effect of large change in Pe_{ax} upon a given reaction. For an analysis of outcomes based upon prior knowledge of the reaction, however, one may prefer to take the precaution necessary to minimize temperature inequalities if heat effects are large.

VALUE FOR R&D PLANNING

The microreactor's greatest immediate value to planners of fluid-bed R&D may lie in experiments revealing effects of axial gas dispersion upon reactions for which little or nothing is known of kinetics or mechanisms.

In the near future, the microreactor could help management avoid expense of a fluid-bed R&D where an economically significant outcome of a reaction is acutely sensitive to axial gas dispersion, and where a fixed bed is an acceptable alternative.

Our hope is that, over time, a body of microreactor data can grow to sufficient size, and represent sufficient variety of reaction type, such that, often, management may use data for a new reaction to justify a decision to build plant around a fluid-bed reactor as soon as laboratory fluid-bed data become available. The goal would be for such a decision to be made with something like the confidence now attached to a decision for plant with a fixed-bed reactor.

ACKNOWLEDGMENTS

National Science Foundation joint industry-university research grant CBT-8620244 supported the work. We are grateful to our collaborators, Frederick J. Krambeck and Amos A. Avidan of Mobil Research & Development Corporation, for their interest, comments, and suggestions. The DuPont Company furnished additional support making possible our study of mass transfer in aerated vibrated beds, and we thank DuPont's Gary Whiting and Peter Compo for their interest and support.

NOTATION

D_{ax} effective axial diffusivity (in fluid bed, m²/s; in vibrated-bed microreactor, cm²/s)
D_{mol} molecular diffusivity (cm²/s)
L length of reaction path (vertical in fluid bed, m; horizontal in microreactor, cm)
Pe_{ax} axial Peclet Number, UL/D_{ax}
S $U/L = 0.1\rho_{cat}\text{WHSV}/(1.27 \times 3{,}600\rho_{gas})$
U superficial gas velocity (vertically upward in fluid bed, m/s; horizontal in microreactor, cm/s)
WHSV weight hourly space velocity, [weight flow of reaction gas per hr]/[weight of catalyst]
ρ_{cat} density of catalyst (compacted)
ρ_{gas} density of reaction gas

LITERATURE CITED

1. Benge, G.G., and A.M. Squires, THIS VOLUME.
2. Squires, A.M., M. Kwauk, and A.A. Avidan, *Science*, **230**(1985):1329-1337.
3. van Deemter, J.J., "Mixing Patterns in Large-Scale Fluidized Beds" in *Fluidization*, J.R. Grace and J.M. Matsen (Eds.), Plenum, New York, 1980, pp. 69-89.
4. Grace, J.R., *Chem. Eng. Sci.*, **45**(1990):1953-1966.
5. Keim, K.-H., J. Maziuk, and A. Tonnesmann, *Erd. Kohle Erdgas*, **37**(1984):558-562.
6. Avidan, A.A., M. Edwards, W. Loeffler, H.-H. Gierlich, N. Thiagarajan, and E. Nitschke, "The Fluid-Bed MTG Process," 21st ACS State-of-the-Art Symposium (Methanol as a Raw Material for Fuels and Chemicals), Marco Island, Florida, June 15-18, 1986.
7. DeMaria, F., J.E. Longfield, and G. Butler, *Ind. Eng. Chem.*, **53**(1961):259-266.
8. Avidan, A.A., *Proc. J. Meet. Chem. Eng. Chem. Ind. Eng. Soc. China Am. Inst. Chem. Eng.* (Beijing, China, Sept. 19-22, 1982), Chemical Industry Press, Beijing, China, 1982, vol. 1, pp. 411-423.
9. Avidan, A.A., and M. Edwards, "Modeling and Scale Up of Mobil's Fluid-Bed MTG Process" in *Fluidization V*, V.K. Ostergaard (Ed.), Engineering Foundation, New York, 1986, pp. 457-464.
10. Avidan, A.A., R.M. Gould, and A.Y. Kam, "Operation of a Circulating fluid-Bed Cold Flow Model of the 100 B/D MTG Demonstration Plant" in *Circulating Fluidized Bed Technology*, P. Basu (Ed.), Pergamon Press, Toronto/New York, 1986, pp. 287-296.
11. Edwards, M., and A.A. Avidan, *Chem. Eng. Sci.*, **41**(1986):829-835.
12. Krambeck, F.J., A.A. Avidan, C.K. Lee, and M.N. Lo, *AIChE J.*, **33**(1987):1727-1734.
13. Bolthrunis, C.O. *Chem. Eng. Progr.* **85** (May, 1989):51-54.
14. Levenspiel, O., *Chemical Reaction Engineering*, Wiley, New York, 2nd edition, 1972.

15. Wen, C.Y., and L.T. Fan, *Models for Flow Systems and Chemical Reactors*, Marcel Dekker, New York, 1975.
16. Raison, C.E., *Mass transfer in aerated vibrated beds*, Masters thesis, Virginia Polytechnic Institute & State University, Blacksburg, 1990.
17. Sprung, B., B. Thomas, Y.A. Liu, and A.M. Squires, "Shallow Vibrated Beds of 'Master Beads'" in *Proceedings* of Fifth International Conference on Fluidization, Elsinore, Denmark. Engineering Foundation, New York City, 1986.
18. Thomas, B., and A.M. Squires, "Vibrated-Bed Microreactors Simulating Catalytic Fluid Beds: A Feasibility Study" in *Fluidization VI*, J.R. Grace, L.W. Shemilt, and M. Bergougnou (Eds.), Engineering Foundation, New York, 1989, pp. 375-382.
19. Socha, R.F., C.T.-W. Chu, and A.A. Avidan, "An Overview of Methanol-to-Olefin [MTO] Research at Mobil: From Inception to Demonstration Plant," 21st ACS State-of-the-Art Symposium (Methanol as a Raw Material for Fuels and Chemicals), Marco Island, Florida, June 15-18, 1986.
20. Tshabalala, S., and A.M. Squires, work in progress at Virginia Polytechnic Institute & State University.

Flow Rate Monitoring and Measurement in Dilute Phase Pneumatic Conveying Using Pressure Fluctuations

Steve Tallon and Clive E. Davies
Industrial Research Limited, PO Box 31-310, Lower Hutt, New Zealand

The static pressure measured at the wall of a pneumatic conveying pipeline contains a fluctuating signal, the nature of which depends on the geometry and flow characteristics of the particular system. These fluctuations travel as sound waves through the pipe and their velocity and intesity are subsequently affected by the flow of the gas and solid phases in the pipe. In this work, estimates are made of the superficial conveying velocity of the gas phase using cross correlation of presssure waves to measure the sound velocity. It is assumed that this velocity is equal to the known velocity of sound in a stationary medium plus the convective velocity of the suspension. A correlation between the variance of the pressure fluctuations and the solids concentration is also developed. Experimental results are presented for a 76mm diameter (3 inch) horizontal conveying system fed by a Rootes type blower and a rotary valve solids feeder. Air flow rates from 0.10 kgs^{-1} to 0.18 kgs^{-1} and solids flow rates up to 0.51 kgs^{-1} were used. The convective velocities calculated by cross correlation showed some sensitivity to sampling parameters. A good correlation was found between the variance of the pressure signal from the Rootes blower and solids concentration.

The pressure measured at the pipe wall in a pneumatic conveying system operating at steady state fluctuates about a mean value. These fluctuations are caused by a variety of factors, including the mechanical action of the air mover, entrance effects at the solids feeder, turbulence effects arising from the motion of the gas and solids in the pipe and around bends, and exit effects associated with the de-entrainment of the solids from the air stream. At any point, the instantaneous pressure is the sum of the mean pressure and the transient pressure waves arriving at that point from all other parts of the system. The form of these fluctuations is dependent on the operating conditions, and changes in these conditions are reflected by changes in the measured pressure signal. As a result, a significant amount of information about the operation of the system can be gained from recording the dynamic pressure signal at the wall of the pipe.

Pressure measurements have been used by many researchers to characterise the operation of gas - solid flow systems. With the advent of digital signal processing, pressure measurements have become a useful tool in the investigation, and on-line analysis, of fluid transport systems. In fluid beds, Kage et al [1] and Dhodapkar and Klinzing [2] use Fourier analysis of pressure signals to identify flow regimes. Tsuji et al [3], and Roy et al [4], use experimental measurements of pressure fluctuations to compare with their correlations and models of the flow in fluidised beds. More recently Daw and Halow [5], and van der Stappen et al [6] have used chaos theory to relate pressure signals to a number of system flow parameters.

In pneumatic conveying systems, Dhodapkar and Klinzing [7] have established relationships between pressure fluctuations, conveying regimes, and other properties such as particle size, pressure transducer type, and system configuration. Their work showed that characteristic power spectral density functions could be established for each conveying regime. Static pressure measurements were found to provide the best spectra for identification of lower concentration regimes, while differential pressure measurements were found better for high concentration regimes.

An approach to actual flow rate measurement in a two phase gas-liquid system was made by Darwich et al [8], using static and differential pressure measurements at the wall of the pipe. They made use of various stochastic features of the pressure signal in both the frequency and amplitude domains. A characteristic map of properties at points on a reference phase flow rate grid was developed, to allow subsequent on-line flow rate estimates using pattern recognition techniques. A number of signal properties were used, but the variance of the signal was found to be the most significant parameter for

---

Industrial Research Limited, Lower Hutt, New Zealand.

resolving flow rates, in terms of a ratio of variation with flow rate to variation at one flow condition.

Other techniques for measuring the mass flow rate of the solids in a pneumatic conveying line have been reported in the literature, but few of these have been developed for industrial use. A device correlating flow rate with the noise in the signal from a capacitance sensor has been described by Beck et al [9], and instruments using this principle have been sold commercially. A more powerful approach which has had some success in the market place obtains a characteristic velocity from the cross correlation of two measurements of capacitance taken a short distance apart, and concentration from the absolute change in capacitance caused by the presence of particles in the pipe. When combined these give a mass flow rate. However, even with this approach there are difficulties, particularly at low product to air mass flow ratios [10].

This paper outlines an approach to the measurement of the mass flow rates of the gas and solid phases from the properties of the naturally occurring pressure signals measured at the wall of the pipe. The method, described below, makes independent measurements of the conveying velocity and the solids concentration from pressure signals recorded at two points along the conveying line. The results presented here are particular to one system, but the techniques should be applicable to flow measurement in most pneumatic conveying systems, and in principle to a range of other flow systems.

Velocity Measurements

The conveying velocity of the gas phase can be estimated by measuring the velocity relative to the pipe of the naturally occurring pressure waves in the system. These pressure waves travel at the speed of sound through the gas - solid suspension and their velocity relative to the pipe is equal to this sound velocity plus the velocity of the suspension [11]. The speed of sound through a stationary suspension can be estimated from known physical parameters, and the velocity of the waves relative to the pipe can be measured by cross correlation of the pressure signals between two points on the pipe. The difference between these two sound velocities gives the convective velocity of the suspension.

This approach has been used by Tallon and Davies [12] to measure the flow rate of air in a single phase pipe flow system. The velocity of sound through the air was estimated from $u_{s0} = \sqrt{\gamma RT}$, where u_{s0} is the sound velocity in a perfect gas, γ is the heat capacity ratio for air, R is the gas constant, and T is temperature. Assuming that the calculated convective velocity was equal to the superficial gas conveying velocity, good agreement was found between these calculated values and independently measured mass flow rates.

The velocity of sound in a two phase flow system is affected by the presence of suspended solids. Several expressions have been developed in the literature relating the sound velocity to voidage and particle parameters. Arastoopour and Gidaspow [13] have developed an expression giving the isothermal sound velocity, from a vertical pipe flow model under near choking conditions, Equation (1). u_s is the velocity of sound in a suspension, ε is the volume fraction of the gas phase.

$$u_s = \sqrt{\varepsilon RT} \quad (1)$$

Roy et al [4] have also developed an isothermal sound velocity, Equation (2), including a density ration term, for sound propagation in fluidised beds. ρ_g is the gas density, ρ_s is the solid particle density.

$$u_s = \sqrt{\frac{RT\rho_g}{\varepsilon[\rho_s(1-\varepsilon)+\rho_g\varepsilon]}} \quad (2)$$

The assumption of isothermal sound propagation is justifiable for solid concentrations approaching those of fluid beds when the solids have a large heat capacity in comparison to the gas, and heat transfer between the gas and solid phase is rapid. This assumption is not valid for very dilute phase pneumatic conveying, and certainly does not apply in the single phase limit $\varepsilon \to 1$.

Gregor and Rumpf [14] have developed an expression for sound velocity in flowing solid - gas suspension. The expression includes terms for particle size, density, sound frequency, and allows for relative velocity and acceleration between the phases. For the range of particle concentrations and sizes used in this work the equations of Gregor and Rumpf [14] predict only very small changes in the sound velocity, but these are still included in the following analysis.

Concentration Correlation

The variance, or standard deviation, of the fluctuating component of the pressure signal is a measure of the intensity of the pressure waves in the system.

There are two main factors which influence the intensity of these waves; changes in the source of the wave, and changes in the attenuation of the wave between the source and the point of measurement. The source of the fluctuation may change in intensity, or frequency, as conveying conditions change, resulting in changes in the variance measured at a point. The attenuation of the waves between the source and the measurement point is through viscous and thermal interaction between the particles, by dispersion, and by friction with the pipe wall. These factors also change as the conveying conditions change, with higher solids loadings generally resulting in greater attenuation and a lower signal variance.

A relationship between the standard deviation of pressure signals and the mass flow rate of solids has been established by Davies et al [15], Hartley et al [16], and extended by Tallon and Davies [12]. Davies et al [15] found qualitatively that, for their system, the low frequency waves associated with the periodic solids feed increased in intensity with increased solids loading, and the higher frequency waves associated with the air feeder were attenuated with higher solids rates.

Tallon and Davies [12] examined the relationship between the variance of the signal, and the gas and solids flow rates for a range of different frequency bands, in the same system described by Davies et al. [15] and used in this work. These results are reproduced in Figures 1 to 4. Low frequency waves, in the 0-20 Hz band, contain signals predominantly associated with the rotary solids feeder, and other rotary valves in the system. The fluctuations are thought to arise from periodic leakage of air through the vanes of the valve, and from periodic acceleration of 'packets' of sand from the feeder [16]. These fluctuations, Figure 1, increased in intensity with increased solids loading. Higher frequency waves, in a 210 to 240 Hz frequency band, Figure 4, contain the other dominant frequency in the system resulting from the motion of the Rootes type blower supplying the air. The variance of this signal was found to be strongest at low, or zero, solids flow and to decrease exponentially with increased loading. The other two frequency bands, Figures 2 and 3, also show a generally decreasing relationship with solids flow, but contain no significant or distinct signals, and are thought to arise from causes such as flow turbulence, and also system noise.

The variance relationships established by Tallon and Davies [12] depend on both the solids and gas phase flow rates. The aim in this work is to find a single expression for these curves which will enable determination of the solids mass flow rate from pressure variance measurements, and an independent measure of the gas phase conveying velocity.

EXPERIMENTAL

The air was supplied by a Rootes type blower, as in Figure 5, and the flow controlled using a bleed valve on the air supply line and an annubar to measure the flow rate (an annubar is a proprietary instrument similar to a pitot tube). The solids were fed through a calibrated orifice in a constant head bin, and through a rotary valve feeder, shown schematically in Figure 6.

The conveying circuit was built from 76mm (three inch) stainless steel tube, in the horizontal plane, and consisted of a 15m straight section, two 90°, 900mm radius bends separated by a 2m straight section, and another 15m straight section of tube. The solids were then conveyed through a short vertical return section, separated through a pair of cyclones and recycled. The exhaust air passed through a bag house filter, and was vented to atmosphere. In all cases the material conveyed was a silica sand with a size distribution as given in Table 1, a bulk density of 1310 kgm^{-3}, and particle density 2480 kgm^{-3}.

Pressures were logged at the pipe wall using solid state differential pressure transducers by Sensym, model 811 SCX05DNC, with a range from 0 to 35 kPa. The transducers were mounted directly onto the pipe wall to minimise the effect on the sound waves of pressure tappings or connecting tubes. The tappings were 4mm diameter holes into which the high pressure port of the transducer was slightly recessed. The transducers were positioned on the top of the horizontal pipe with the low pressure port left open to the atmosphere.

Analogue to digital conversion was by a data acquisition unit, model 901A, from Strobes Engineering

Table 1 - Sand particle size distribution

Mesh Size (μm)	355	250	150	106	53
Weight % Undersize	98.8	84.2	35.3	15.2	3.4

Ltd, New Zealand, capable of sampling rates up to 100,000 Hz. Pressure samples 16384 points long were logged simultaneously for points 10.75m and 13.75m, and for points 10.75m and 11.75m, from the centerline of the solids feeder (i.e a few meters before the first bend). This allowed cross correlation over an axial length of 1m and 3m. Sampling frequencies of 5000 Hz and 100,000 Hz were used for a range of air flow rates from 0.10 to 0.18 kgs^{-1} (conveying velocities circa 20 to 35 ms^{-1}) and for sand flow rates up to 0.51 kgs^{-1}. Conveying air temperatures at the pressure tappings were measured using a thermocouple for each flow setting.

ANALYSIS

The velocity of the sound waves through the pipe was estimated from the time lag at the peak of the cross correlation function, as defined by Bendat and Piersol [17], between pressure recordings at the two spacings along the pipe, 1m and 3m. The velocity of sound through the suspension was estimated, as an initial approximation, from the velocity of sound in pure air, $\sqrt{\gamma RT}$, and also by the equations of Gregor and Rumpf [14]. The mass mean particle diameter and the dominant signal frequency, 223 Hz, are used in the equations. As in Gregor and Rumpf, the sound waves are considered to be travelling through the continuous gas phase, so that the convective velocity of the wave is equal to the gas phase velocity. The estimated mass flow rate of air was calculated from the difference between the two sound velocities multiplied by the cross sectional area of the pipe, and the average air density at the measured temperature and mean pressure of the recording. The small correction for voidage, greater than 0.997 in this work, is ignored for simplicity by assuming the gas phase velocity is equal to the superficial gas velocity.

The attenuation of a sound wave passing through a suspension can be described by Lambert-Beer's law, as in Equation (3). For example see Iinoya et al [18]. I_0 is the initial intensity, I_1 is the intensity at the point of interest, l is the path length through which the wave passes, and α is an attenuation or extinction coefficient. A similar relationship is also expected to apply to the variance of pressure signals in the pipe. In a pneumatic conveying system the pressure measured at a point represents the sum of the pressure waves arriving from all disturbances in the pipe. In this work however, the analysis is concentrated on a narrow frequency band, 210 to 240 Hz, which consists primarily of the signal from the air blower.

$$I_1 = I_0 e^{-\alpha l} \quad (3)$$

The attenuation coefficient represents attenuation due to the viscous and thermal interaction of the particles, scattering, dispersion of the waves due to turbulence, and interaction with the pipe wall. This coefficient is a function of the particle and wall properties, particle flow rates, and the solids concentration. The source intensity, I_0, of the wave is subject to change with changes in the operating condition and will affect any correlation of variance with flow conditions. The path length through which the sound wave passes also depends on the conveying conditions. If the pressure is recorded at a fixed point on the pipe then the distance the wave travels through the suspension depends on the convective velocity of the air. The distance travelled through the suspension will be less than the distance relative to the pipe when the wave speed is increased by the air flow.

RESULTS AND DISCUSSION

Cross Correlation

The air flow rates calculated using the sound velocity in pure air are shown against the values measured by the annubar in Figures 7 and 8 for correlation over distances of 1m and 3m respectively. For the 1m correlation distance, Figure 7, the calculated values give higher values than measured by the annubar. Over the 3m distance, Figure 8, the calculated values are lower than the annubar values, but in general the expected trends can be observed.

The discrepancy between calculated and measured mass flow rates is typically up to 0.04 kgs^{-1}, or a velocity difference of 7 ms^{-1}. This discrepancy can be accounted for by a number of uncertainties associated with the calculation. Due to the discrete number of points taken in the digital sample the velocity of the sound wave can only be resolved to within about 2.5ms^{-1} for a distance of 1m and 100,000 Hz sampling speed. The estimate of the sound velocity in the air is very sensitive to the temperature measurement, and as the convective velocity is calculated from the difference of two larger velocities, this sensitivity is increased. Small uncertainties in the distance between correlation points are also amplified by the difference calculation.

Other areas of uncertainty lie in the assumed models for the convection of sound waves in the pipe. Braun et al [19] describe the development of a capacitive cross correlation device for use in pneumatic conveyors and observe deviations between the cross correlation transit velocity and true mean velocities of up to 10% due to model errors. Any resonance in the pipe due to reflection of sound waves will also introduce errors, and scatter, in the cross correlation measurements due to the phase shifts associated with a system containing standing waves.

For both the 1m and 3m cases the higher solids flow rates give lower calculated air flow rates. This is consistent with a reduction in the speed of sound due to the presence of solids in suspension. Figure 9 shows correlation over the 3m length using the equations of Gregor and Rumpf [14] to estimate the sound velocity in the suspension. The result shows some improvement to the grouping of the different solids flow rates, although there is still uncertainty in the use of equations developed for single frequency sound in a supension of monosized particles. Further investigation is also needed into the effects of sampling parameters, measurement uncertainty, and model errors, to improve this correlation.

Variance Correlation

Figure 10 shows the pressure variance in the 210 to 240 Hz frequency range plotted on a logarithmic scale against the sand flow rate. This frequency band predominantly contains the fundamental signal from the Rootes blower. The variance is shown at a point just before the feeder, and at a point 10.75 m from the feeder. The variance at the point along the pipe shows a roughly linear relationship on a log scale except for some deviation at low and zero solids flow rates. The variance from before the feeder is relatively constant with air and sand rates for higher solids loadings, but again at lower solids flows the variance shows some variation. This suggests that the strength of the signal from the Rootes type blower is reasonably constant except at lower solids rates when the back pressure to the blower is small.

If the attenuation model of Equation (3) is assumed then the pressure variance should be a function of the conveying rates of the solid and gas phases, and of the solids concentration. If the attenuation is due mainly to the thermal and viscous interaction of the gas and solid particles then, for constant particle properties, the attenuation is expected to be a function purely of the concentration of solids. Figure 11 shows the log of the pressure variance plotted against the solids volume concentration. The volume concentration of solids was calculated from the known gas and solid mass flow rates, assuming a slip velocity of 3.0 ms^{-1}. The concentration was calculated using the air flow rates measured by the annubar, rather than the cross correlation values. The resulting figure shows a strong correlation, with the main deviations occurring at the high and low ends of the concentration range.

Deviations at low concentrations are expected as changes in the strength of the blower signal occur at lower concentrations, Figure 10. At high solids loading, the signal from the blower is almost nonexistent at the point of measurement (variance < 10 Pa^2). Consequently the effects of other signals, noise, and even the effects of digital processing such as filtering, become significant and the relationship breaks down.

The variance is expected to have some dependence on the individual phase flow rates. The effective path length, as noted above, depends on the conveying velocity and in this work varies by about 5% between the lowest and the highest velocity. Attenuation of the waves due to dispersion by the flow turbulence is expected to be not just a function of the concentration as the distribution of solids is dependent on the conveying velocities. Different spatial distributions of the solid particles, and changes in the flow turbulence, will have different attenuating effects on the pressure waves.

As a simple initial approach to fitting the data and accounting for variations with different conditions a linear model was assumed as in Equation 4, where C_1 to C_4 are constants, w_g is the gas mass flow rate, w_s is the solids mass flow rate, and c is the solids volume concentration. A least squares error method was used to evaluate the constant parameters over the range of flow rates used in this work. Estimates of the concentration using this correlation, and the air mass flow rates recorded by the annubar, are given in Figure 12. The fit is an improvement over the concentration regression, Figure 11, particularly at lower concentration, with typical deviations in estimated concentration values less than 4x10^{-5} (vol/vol)

$$\log(\sigma^2) = C_1 + C_2 w_g + C_3 w_s + C_4 c \qquad (4)$$

The solids volume concentration is a function of the mass flow rates of the gas and solid phases, and of the difference between the mean solids velocity and the mean gas velocity (i.e. the slip velocity). Thus the solids mass flow rate can be estimated using Equation (4) given the mass flow rate of the gas phase, the pressure signal variance, and an assumed value of the slip velocity.

CONCLUSIONS

The conveying velocity can be estimated by measuring the speed of natural sound waves travelling through the pipe. Results using this approach were found to be realistic but sensitive to the sampling parameters, small measurement uncertainties, and to errors in the assumptions. The variation in calculated values with changes in solids flow rate was small for the conditions used in this work, but was largely accounted for by correlations predicting the sound velocity in a suspension.

The variance of a pressure signal generated by the Rootes blower was found to be a strong function of the solids volume concentration. A linear fit of experimental data to solids mass flow rate, gas mass flow rate, and solids volume concentration, yields a good correlation. Given an independently measured air flow rate, such as by cross correlation of sound waves, the mass flow rate of solids could be determined.

ACKNOWLEDGMENT

We would like to thank one of the reviewers for comments on the effects that the presence of solid particles can have on the sonic velocity in suspensions.

NOTATION

c	Volume concentration of solids
$C_{1\,to\,4}$	Linear regression constants
I_0	Intensity at reference point
I_1	Intensity at measurement point
ℓ	Path length
R	Gas constant
T	Absolute temperature
u_s	Sound velocity in a suspension
u_{s0}	Sound velocity in a perfect gas
w_g	Gas mass flow rate
w_s	Solids mass flow rate

Greek Letters

α	Attenuation coefficient
ε	Voidage
σ^2	Variance
ρ_g	Gas Density
ρ_s	Solid Particle density
γ	Heat capacity ratio

LITERATURE CITED

1. Kage, H., Iwasaki, N., Matsuno, Y. *AIChE Symposium Series*, **89**, p.184, (1993)

2. Dhodapkar, S., Klinzing, G., *AIChE Symposium Series*, **89**, p.170, (1993)

3. Tsuji, Y., Kawaguchi, T., Tanaka, T., *Powder Tech.*, **77**, p. 79, (1993)

4. Roy, R., Davidson, J. F., Tuponogov, V. G., *Chemical Engineering Science*, **45**, p.1, (1990)

5. Daw, C. S., Halow, J. S., *AIChE Symposium Series*, **89**, p.103, (1993)

6. van der Stappen, M. L. M., Schouten, J. C., van den Bleek, C. M., *AIChE Symposium Series*, **89**, p.91, (1993)

7. Dhodapkar, S. V., Klinzing, G. E., *Powder Tech.*, **74**, p. 179, (1993)

8. Darwich, T. D., Toral, H., Archer, J. S., *SPE Production Engineering*, p.265, August (1991)

9. Beck, M. S., Green, R. G., Plaskowski, A. B., Stott, A. L., *Meas. Sci. Tech.*, **1**, p.561, (1990)

10. Woodhead, S. R., Barnes, R. N., Reed, A. R., *Third Int. Conf. on Bulk Materials Storage Handling and Transportation*, Newcastle, Australia, 27 to 29 June, p.289, (1989)

11. Munjal, M. L., *Acoustics of ducts and mufflers*, John Wiley and Sons, (1987)

12. Tallon, S., Davies, C. E., *The use of Pressure Fluctuations to Characterise Gas and Solids Flow Rates in Dilute Phase Pneumatic Conveying*, Presented at Chemeca '94, 25th to 28th September, Perth, Australia (1994)

13. Arastoopour, H., Gidaspow, D., *Ind. Eng. Chem. Fundam.*, **18**, 2, (1979)

14. Gregor, W., Rumpf, H., *Powder Tech.*, **15**, p. 43, (1976)

15. Davies, C. E., Gunawan, I., Hartley, M., *Correlation of Pipeline Pressure Fluctuations with Solids Flowrate in Dilute Phase Pneumatic Conveying*,

Presented at Chemeca '91, Australia, 18th to 20th September, Newcastle, p.833 (1991)

16. Hartley, M., Davies, C. E., Gunawan, I., Fenton, K., *Flowrate-Pressure Fluctuation Correlation in Dilute Phase Pneumatic Conveying: Effect of System and Operating variables*, Presented at Chemeca '92, 27th to 30th September, Canberra, Australia, p.591 (1992)

17. Bendat, J.S., Piersol, A.G., *Random data - Analysis and measurement procedures*, John Wiley & Sons, (1986).

18. Iinoya, K., Masuda, H., Watanabe, K., *Powder and Bulk Solids Handling Processes*, Marcel Dekker, (1988)

19. Braun, H., Füg, M., Schneider, G., *Chem. Eng. Technol.*, **10**, p.353, (1987)

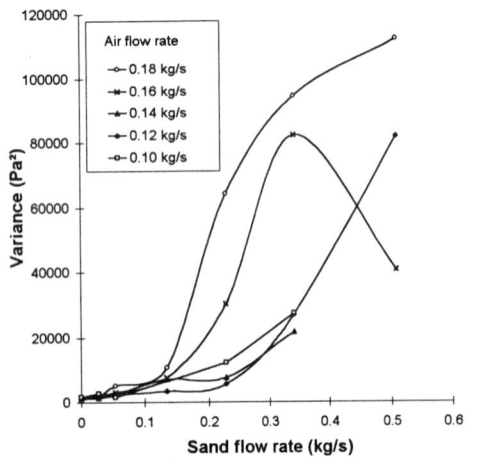

Figure 1 Sand flow rate vs pressure variance, with air flow rate as a parameter. 0 to 20 Hz frequency band.

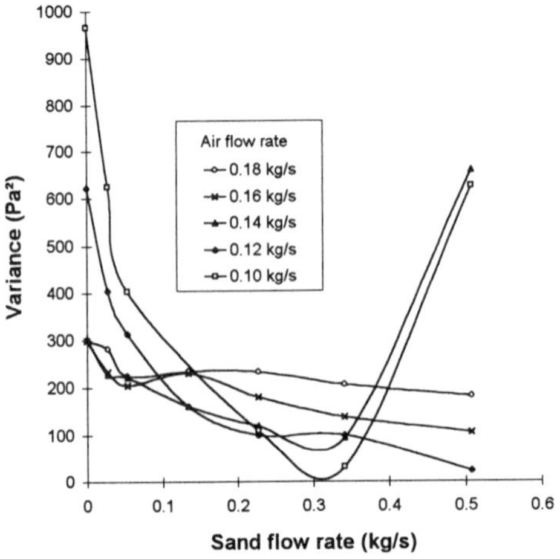

Figure 2 Sand flow rate vs pressure variance, with air flow rate as a parameter. 20 to 90 Hz frequency band.

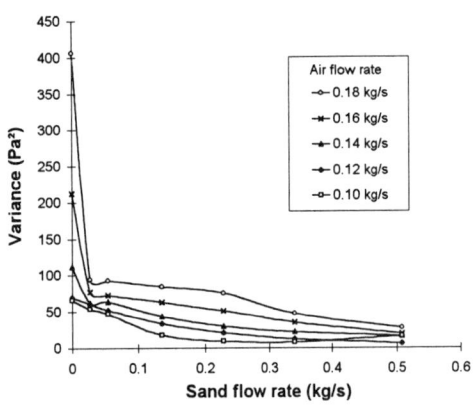
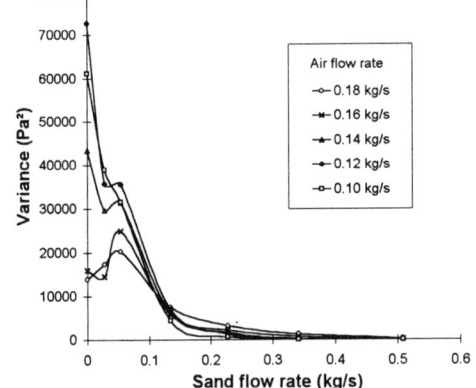

Figure 3 Sand flow rate vs pressure variance, with air flow rate as a parameter. 90 to 210 Hz frequency band.

Figure 4 Sand flow rate vs pressure variance, with air flow rate as a parameter. 210 to 240 Hz frequency band.

From Tallon and Davies [12]
Pressure variance in different frequency bands for 150µm diameter sand conveyed in air in a 3" diameter horizontal pipeline. Air supplied by Rootes blower, and sand by rotary feeder. Samples taken at 5000 Hz at a point 10.75m after the feeder.

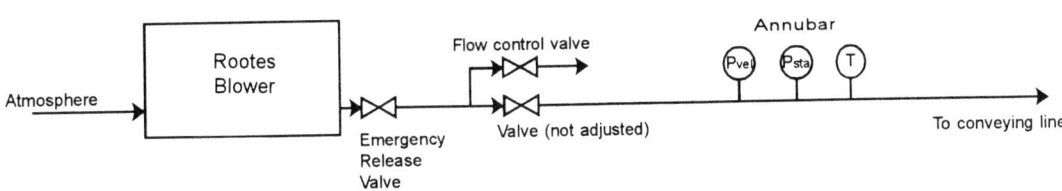

Figure 5 - Air feed layout

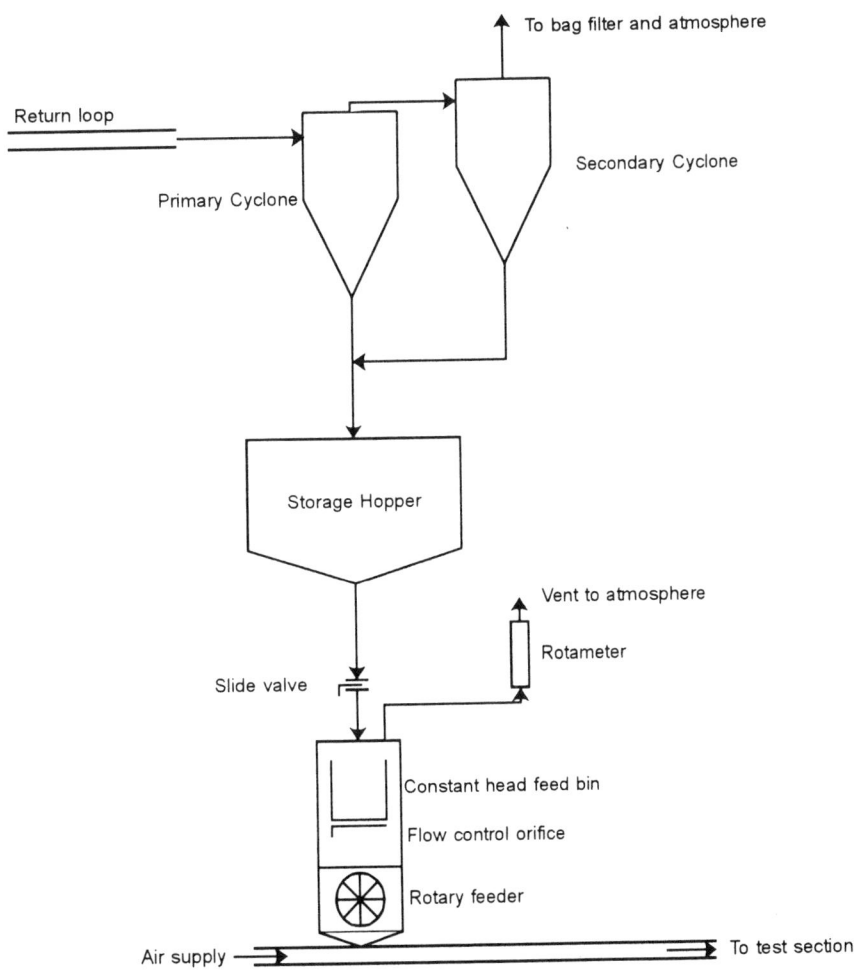

Figure 6 - Solids feed layout

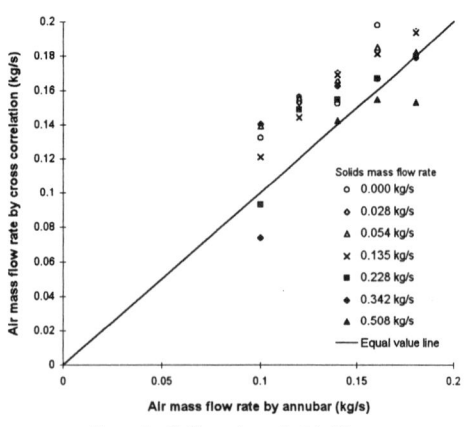

Figure 7 - Air flow rates calculated by cross correlation of sound waves over 1m distance

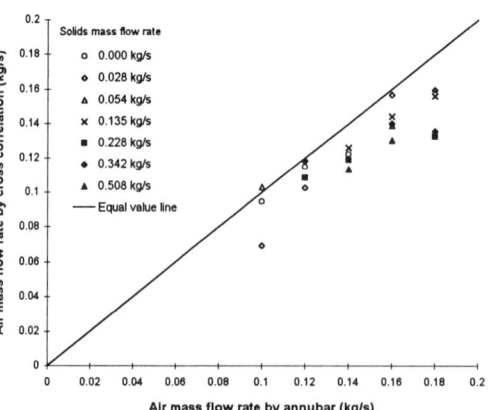

Figure 8 - Air flow rates calculated by cross correlation of sound waves over 3m distance

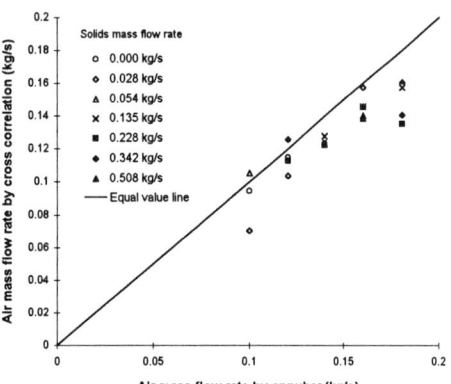

Figure 9 - Air flow rates calculated by cross correlation of sound waves over 3m distance. Sound velocity from Gregor and Rumpf [14]

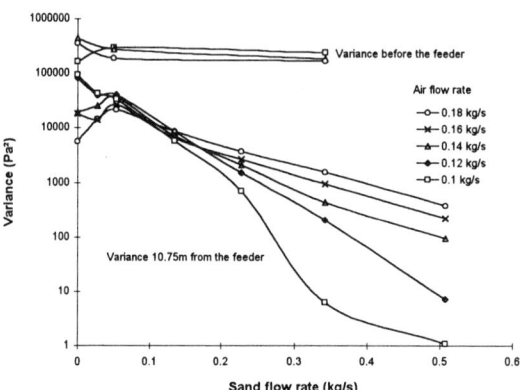

Figure 10 - Variance of 210 to 240 Hz frequencies, against sand flow rate showing variance at a point, and variance before the feeder

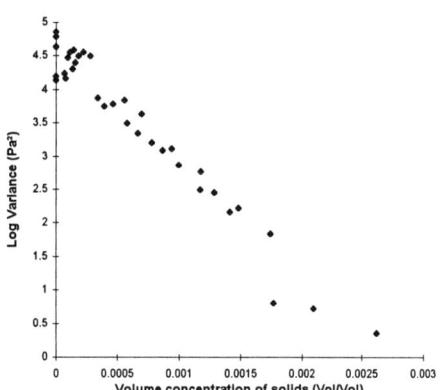

Figure 11 - Variance against solids volume concentration

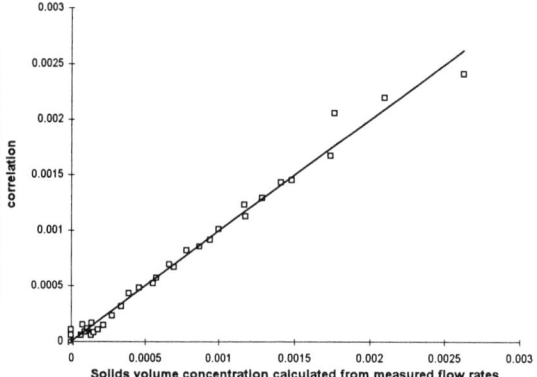

Figure 12 - Solids concentration from linear fit to pressure variance. Air flow rates 0.10 to 0.18 kg/s, sand flow rates 0 to 0.51 kg/s

Simultaneous *In-Situ* Determination of Particle Loadings and Velocities in a Gaseous Medium

Ray Cocco, John Cleveland, Rich Harner and Ray Chrisman
The Dow Chemical Company, Midland, Michigan 48674

The in-situ determination of solids loading and velocity in dynamic gas-solid systems has traditionally been a complicated problem. Standard techniques do not generally provide enough information, are often plagued by low sensitivity, and are rarely flexible enough to be employed in-situ. These shortcomings have hindered efforts to model the kinetic and hydrodynamic behaviors of fluidized beds and riser reactors for many years. In view of this, we have developed an in-situ technique, called Backscattered Particle Imaging or BPI, which allows for the simultaneous measurement of solids loading and velocity. The technique is based on backscattered light from a fiber optic probe. Reflected light from passing particles provides real time waveforms of particle hydrodynamics. From the particle waveforms, a numerical algorithm is used to determine solids loading (volumetric solids concentration) and solids velocity.

Measuring solids loadings and velocities in dynamic gas-solid systems has traditionally been a complicated problem. Although there is a wide range of techniques [1-7], varying from simple pressure devices to the more complex laser Doppler anemometry, few of these techniques offer simultaneous detection of solids loading and velocity in reactive environments. In addition most of these techniques require elaborate equipment implementation and/or are capital intensive. In view of this, Dow Chemical has developed a "Backscattered Particle Imaging" (BPI) probe for low-cost *in-situ* determination of particle concentration and velocity [8].

The BPI probe utilizes backscattered light from a passing particle to infer the solids loading and velocity. Resolution of particle velocities for relatively small diameters can be obtained by converging the emitter and receiver fiber such that only a small detection volume exists [9-10]. Lyons, et. al. [10] applied this approach for measuring particle flow rates in multi-phase media, and proposed that the 50% autocorrelation time is proportional to the transit time of a single particle through the field of view of the probe.

We found, that although this technique worked well for low solids loadings in a gas phase, it was limited to solids loadings of less than 10%. However, by using a normalized first derivative instead of an autocorrelation, we were able to expand on Lyons' method such that velocity measurements can be obtained in gas-solid streams with solids loadings of 50%. Experimental results, using Group A and B [11] alumina particles, have shown that this technique is highly versatile with applications in lean and dense phase systems. The probe was also effective in detecting slugs (i.e., size and frequency) in a riser. As a result, the BPI probe can be used in gas-particle systems such as risers, pneumatic conveyers, fluidized beds and cyclone diplegs.

EXPERIMENTAL

Probe Design

A schematic of the BPI probe design is shown in Figure 1. The illuminating source was a 200 μm diameter fiber bent 15° from the normal axis of the probe face. The collecting fiber was a 400 μm or 200 μm diameter fiber also bent 15° from the normal axis. The illuminating and collecting fibers converge toward each other and were held in place with an adhesive. Optimum sensitivity was found when the collecting fiber being downstream of the illuminating fiber. The diameter of the probe housing was 0.79 cm (5/16").

The illuminating fiber was connected to an LED operating at 850 nm. The collecting fiber was connected to a silicon photodiode. In this work, the LED was operated at a constant 100 ma and the amplifier at 10^7 I/V gain and

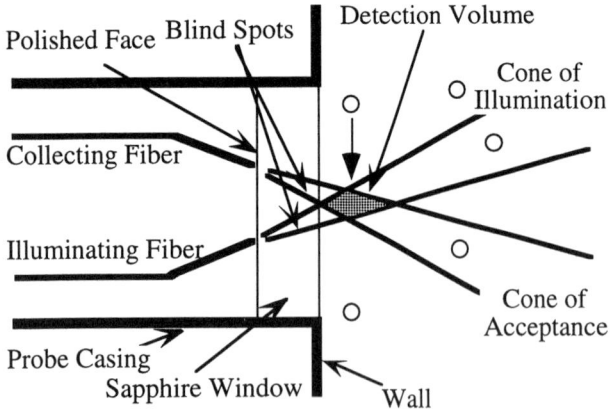

Figure 1. Schematic of the BPI probe operation.

10 kHz bandwidth. The amplified signal was fed into a CIO-AD16 A/D card (ComputerBoards, Inc., Mansfield, MA) and a Compact 386e computer. Data acquisition was done using QuickBasic 4.5 (Microsoft).

Since the cladding and protective coating diameter of the collecting and illuminating fibers is not negligible, the principal rays intersect ~2 mm beyond the polished face. If the probe were used in this manner, a significant error would be introduced. The backscattered light signal would increase with concentration until a significant number of particles begin to occupy the "blind spots" between the polished face and the detection zone. At this point the backscattered intensity would decrease because the illuminating light traveling to detection zone is attenuated [12]. This problem can be avoided by placing a window in front of the prolished face. By eliminating this "dead space," the backscattered intensity increases monotonically with concentration.

The probe was constructed for high temperature applications using materials with similar thermal expansion coefficients as that of the silica fibers. As shown in figure 2, an Invar insert was used to support the fibers into the 15° converging configuration. An adhesive was used to support the fibers in the invar insert. At the face, the fibers were cleaved and polished to a flat surface. The insert was retained in a titanium alloy Gr7/stainless steel body with a stainless steel spring. This design allows the probe to be serviced without breaking the process seal. The sapphire window was brazed to the titanium alloy body by 3E Laboratories (Montgomeryville, PA). An adhesive was used to bind the stainless steel and titanium alloy together to form the body.

Fibers were mounted in a triad geometry as shown in figure 2. Of a total of three available fibers, one was used as a source; another was used for as a receiver. The third fiber was reserved as a spare in case one of the other fractured. The core size of the optical fibers selected was dependent on the particle diameters under investigation. For the larger Group B powder, 400 µm diameter silica fibers were used for the collecting fibers, whereas for the Group A powder, 200 µm diameter fibers were used. Details for this criteria are discussed below.

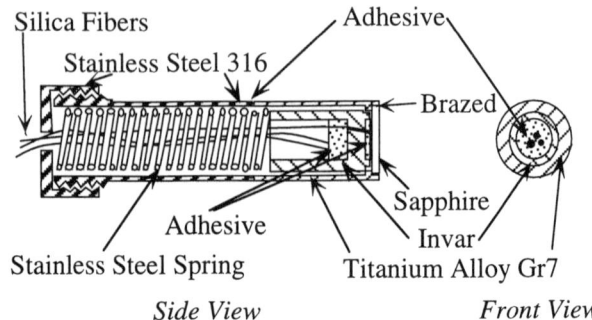

Figure 2. Schematic of the BPI probe construction.

Equipment

The BPI probe was evaluated in two cold flow units: a three-meter particle drop tube and a continuous riser/fluidized bed unit. Schematics of these units are shown in figure 3. Measurements for low solids loadings (>95% voidage) were obtained in the three-meter particle drop tube. The probe was located sufficiently far from the hopper to insure well-developed flow. The distance was checked by determining the acceleration and terminal velocity for the particles. Loadings were determined by terminating the solids flow via the gate valve and measuring the solids hold-up in the tube. Particle velocities were calculated using the expression of Haider and Levenspiel [13].

Medium solids loadings (65 to 95% voidage) measurements were obtained in the riser portion of the riser/fluidized bed unit. The riser dimensions were one meter in length with an inner diameter of 0.9 cm and outer diameter of 1.27 cm. The probe was located in two places, both sufficiently up-stream of the acceleration region. Solids loadings or voidages were determined by terminating gas flow to the riser and fluidized bed. Solids remaining in the riser corresponded to the solids hold up during operation.

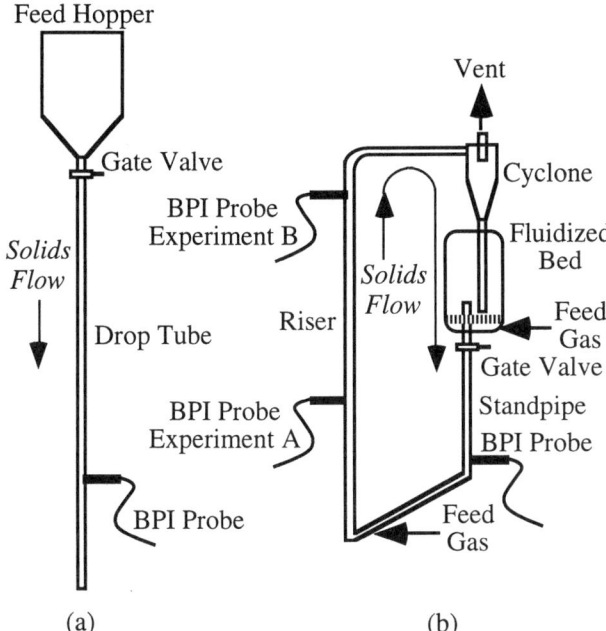

Figure 3. Schematic of the (a) particle drop tube and (b) riser/fluidized bed unit.

Solids velocity measurements were obtained by measuring the fill rate in the cyclone when the fluidized bed was not in operation. The termination of feed gas to the fluidized bed prevented solids flow through the dipleg. Solids loading in the riser was metered using a gate valve in the fluidized bed standpipe. Care was taken to insure that the solid feed to the riser was unaffected during the filling of the cyclone

High solids loadings (40 to 65% voidage) measurements were obtained in the standpipe of the fluidized bed. Similar experiments could also be performed in the cyclone dipleg. Solids velocity measurements were obtained by clocking colored particles used as tracers in the standpipe. Confirmation of this technique was obtained by timing fill rates of the standpipe emptying into a beaker instead of the riser.

In all cases the probe was mounted flush to the wall of the drop tube, riser or standpipe. The capability of moving the probe into the flow stream was available, however, it was not pursued for this study.

Two types of powders were used in this study. The Group A powder was a standard alumina-based FCC catalyst with a mean diameter of 76 μm. The Group B powder was also alumina-based catalyst with a mean diameter of 600 μm. Both catalysts were obtained from Englehardt.

RESULTS AND DISCUSSION

In-Situ Measuring of Solids Loading

The BPI probe mean intensities plotted against the measured solids loadings for particle flow in lean and dense phase systems are shown in Figure 4. The mean intensity of the BPI probe varied linearly with the measured solids loading. The relatively large variation in data at high solids loading was due to difficulties in the experimental determination of solids loading in the standpipe and not due to variations in the probe's response. The error bars correspond to variations among measurements made in an experiment. At low solids loadings, the error bars are smaller than the data points, and cannot be observed. Each data point for the BPI mean intensity is the result of the collection of 100 scans with each scan consisting of 1000 points.

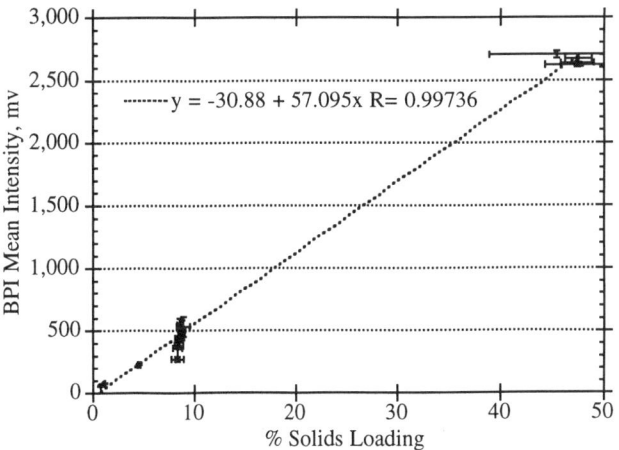

Figure 4. BPI probe mean intensity compared to measured % solid loadings for particles with a mean diameter of 600 μm (Group B). A sampling frequency of 10 kHz was used.

The accuracy of the BPI probe is dependent on the sampling frequency. Higher sampling frequencies will provide a better peak resolution in the waveform. However, the resulting shorter collection time may not provide enough data (or peaks) to insure a statistically significant BPI average intensity. For instance, the "unresolved" data in the 8 to 10% solids loadings of Figure 4 is due to a low sampling frequency (10 kHz). However, this sampling frequency was needed in order to examine a wide range of

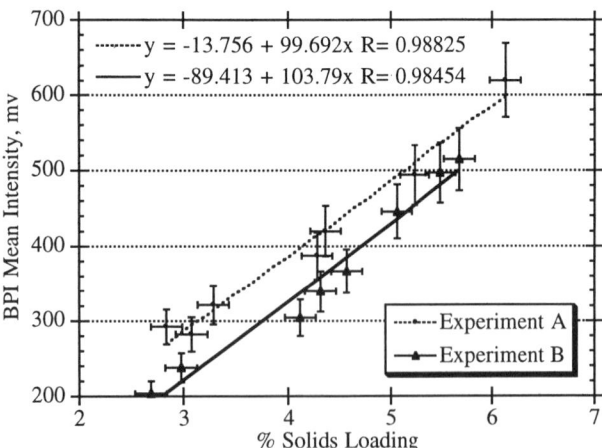

Figure 5. BPI probe mean intensity, for Group B powder, collected for two independent experiments in a riser using a sampling frequency of 20 kHz.

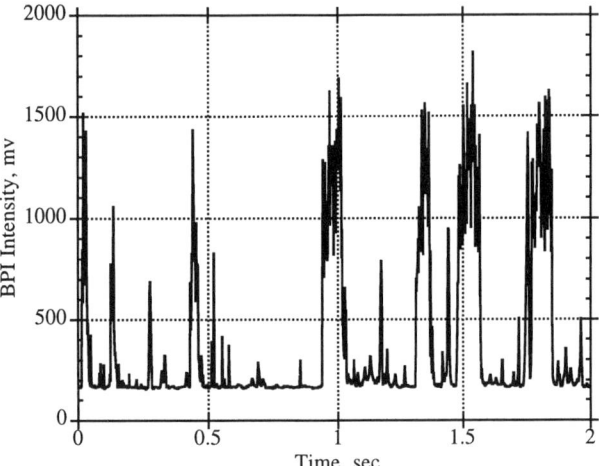

Figure 6. BPI intensity for solids flow in a riser operating near the choking velocity.

solids loadings (i.e., 0 to 50%). In examining narrower ranges of loadings, a higher sampling frequency (20 kHz) can better resolve the data as shown in the separate experiments of Figure 5.

Figure 5 also demonstrates the reproducibility of the BPI mean intensity. Two sets of data were collected (Experiments A and B) under similar riser conditions with the exception that the BPI probe was located further downstream for Experiment B. As shown in Figure 5, the BPI probe was sensitive to solids loading changes of ~1% with the slope for the BPI response for each experiment being nearly equal. Variation in the y-intercept is due to error in baseline subtraction.

Examining the raw data from the BPI probe provides imaging of the gas-solids hydrodynamics. Figure 6 shows the backscattered intensities from the probe with respect to time for solids flow in the riser. With the riser operating near the choking velocity, the BPI probe was capable of resolving both particle and slugging events. Thus, this technique can also be used to infer slugging frequencies and size in a reactive environment.

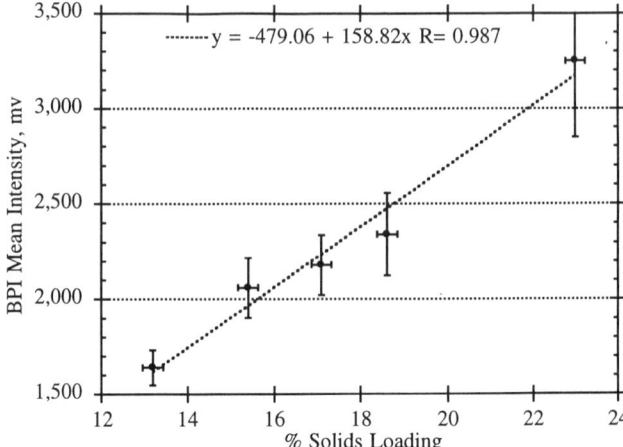

Figure 7. BPI prove mean intensity, for Group A powder, in a riser using a sampling frequency of 30 kHz.

Group A powders were also examined with the BPI probe in the riser. As shown in figure 7, the mean intensity from the BPI probe also increased linearly with increasing solids loadings. Here, 200 μm diameter fibers were used for the Group A powder instead of the 400 μm diameter particles used for the Group B powder. Accuracy in determining the solids loadings for the Group A powder was within 2%.

In-Situ Measuring of Solids Velocity

Determining the velocity of a single particle with backscattered light poses little complications. The width of a single, Gaussian shaped peak [14] in the waveform is indicative of the particle travel time through the detection volume. Multiple peaks, however, complicate the problem. Lyons [10] used autocorrelation to correlate particle velocities to the resulting waveform for particle flow in a liquid medium. This technique also worked well for particles in a gaseous medium as long as solids loadings are low. At higher solids loadings, autocorrelation failed.

The derivative of the waveform, however, is also indicative of the particle velocity. For instance, backscattered light [14] from a particle passing through a parallel set of illuminating and collecting fibers can be represented as

$$I_{x,y,z} = A \exp\left(\frac{-t^2}{2\sigma}\right) \quad (1)$$

where $A = \dfrac{\pi R^2}{z^4 \tan^2 \theta} \exp\left(\dfrac{-2[y^2 - s^2]}{z^2 \tan^2 \theta}\right)$ and $\sigma = \dfrac{z \tan \theta}{v} = \dfrac{k'}{v}$

The parameters x,y,z, θ and s are defined in figure 8 and v is the velocity of the particle. The coefficient A is the amplitude and σ^2 is the variance of the resulting Gaussian peak. k' is some constant.

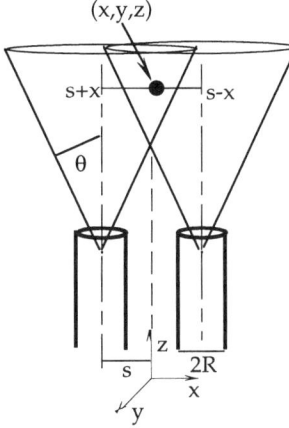

Figure 8. Schematic of backscattered light from a particle using two parallel fibers.

Taking the derivative of equation 1 yields the expression

$$\frac{dI_{x,y,z}}{dt} = \frac{-tA}{\sigma^2} \exp\left(\frac{-t^2}{2\sigma}\right) \quad (2)$$

The maximum derivative is the most convenient for use in the analysis as it can be easily found at the peak inflection point. Taking the second derivative of equation 2 and setting it to zero results in the expression

$$\frac{d^2 I_{x,y,z}}{dt^2} = \frac{1}{\sigma^2}\left(\frac{t^2}{\sigma^2} - 1\right) \exp\left(\frac{-t^2}{2\sigma}\right) = 0 \quad (3)$$

Equation 3 is satisfied when t=σ (at the inflection point). Thus, the maximum derivative, given by the expression

$$\left.\frac{dI_{x,y,z}}{dt}\right|_{max} = \left.\frac{dI_{x,y,z}}{dt}\right|_{t=\sigma} = -0.61 \frac{A}{\sigma} = \frac{k''}{\sigma} = \frac{k''}{k'} v \quad (4)$$

is proportional to the particle velocity (speed maybe more appropriate as the probe does not measure direction).

A comparison of the maximum derivative of the backscattered waveform with the measured solids velocity demonstrates that a solids concentration dependency exists. Normalizing the absolute maximum derivative with respect to the concentration raised to the one-third power, $(ABS(dI/dn)|_{max})/I^{0.33}$, results in a linear relationship with experimentally determined solids velocity over a wide range of solids loadings, as shown in Figure 9 for Group B powder. This normalized response is referred to as the velocity modulus. The mean velocity modulus is the results of averaging the maximum derivative of 100 scans divided by the mean BPI intensity to the one-third power. The error bars in the y-direction of figure 9 is the standard deviation of the mean velocity modulus.

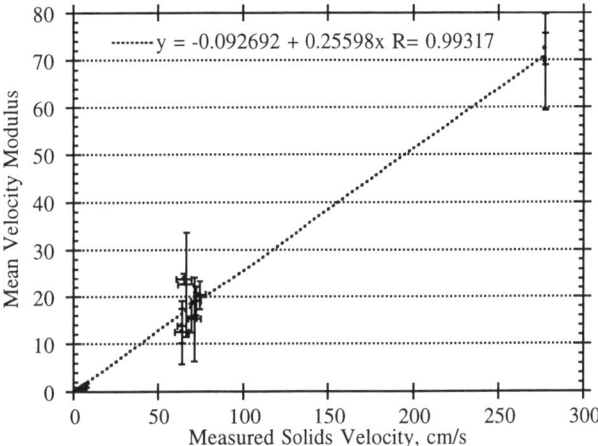

Figure 9. Mean velocity modulus versus the measured solids velocity data for particles with a mean diameter of 600 µm (Group B). All data collected with a sampling frequency of 10 kHz.

Slow particle velocity measurements in figure 9 were obtained in the fluidized bed standpipe. Intermediate velocities (50 to 280 cm/s) were obtained in a riser and calculated from measured average solids flow rates and loadings. High particle velocities were obtained in the particle drop tube. The control measurements (values to be compared to the BPI probe) were typically performed at least ten times to increase statistical significance. The x-error bars represent the variations in these measurements. The x-error bars for the low velocity measurements (0 to 10 cm/s) are masked by the data points. There are no x-error bars for the high velocity measurements (250 to 300 cm/s) as that was deterimined from terminal velocity calculations.

As with the solid loading measurements, the choice of the sampling frequency will effect the accuracy of the velocity measurements. This is demonstrated in Figure 10 where a faster sample frequency of 20 kHz provided better

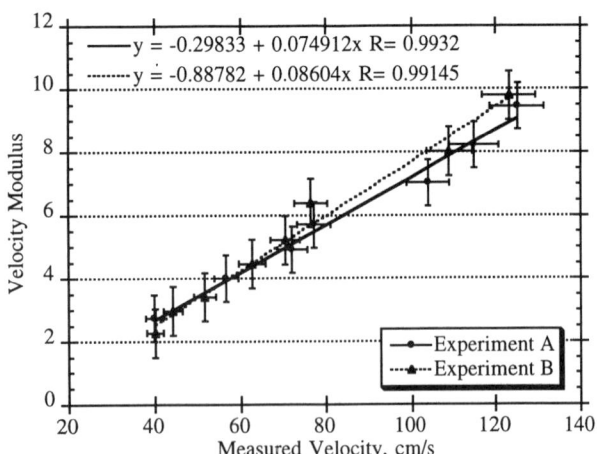

Figure 10. Velocity modulus compared to measured data for Group B powder in the riser at a sample frequency of 20 kHz.

particle resolution in the flow regime of 20 to 140 cm/s than that obtained with 10 kHz (figure 9).

The choice of the correct sampling frequency is a trade off between accuracy and range of analysis. If the sampling frequency is low, a wide range of velocities can be examined, but error is increased as there may not be enough sample points to adequately describe the waveform. If the sampling frequency is high, the total number of peaks collected may be too small to accurately measure the dynamics of the gas-solid system. Adding more sample points can increase the size of this waveform, but at the expense of longer processing times. To insure accurate analysis, each waveform should contain several peak events.

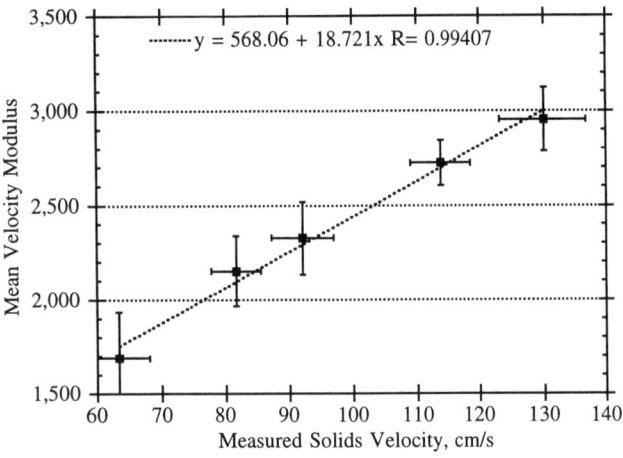

Figure 11. Mean velocity modulus compared to masured data for Group A powder in the riser at a sampling frequency of 30 kHz.

A similar linear relationship was observed for the mean velocity modulus and measured velocities for Group A powders, as shown in figure 11. These results were dependent on reducing the collecting fiber diameter from 400 μm, used for the Group B powders, to 200 μm. The 400 μm diameter fiber provided a detection volume too large to resolve individual peak events for high solids loading systems. Decreasing the fiber diameter and increasing the sampling frequency to 30 kHz provided sufficient peak resolution in the waveform to determine derivatives (i.e., the solids velocity of individual peak events).

Limitations

As with most techniques, there are some limitations that need to be considered. First, the mean BPI intensity and velocity modulus should contain enough points to have statistical significance. Since the surfaces of the detection volume are not normal to the particle trajectory (figure 8) and detection efficiency varies throughout the volume, two particles having the same velocity but entering the detection volume at different locations may have different peak intensities and derivatives. By taking the mean of the intensity and velocity modulus over a large collection of scans, variation due to the entry point of each particle in the detection volume is averaged out, thus a linear relationship with measured data is obtained. The variation of particle trajectories will only be observed as part of the variance.

Second, agglomeration of particles, such as in slugging (figure 6), causes the velocity modulus to be sensitive only to the first and last particle in the slug, where higher maximum absolute derivatives are observed. This is due to the sampling frequency not being high enough to resolve particle velocities in a more dense loading. Higher sampling frequencies will provide the needed increase in resolution, but the range of particle velocities being measured becomes more limited. If we assume that the velocity of the first and last particle are indicative of all other particles in the slug, then the error in the mean velocity modulus due to slugging is small.

Finally, using the mean intensity of the BPI probe to determine solids loadings assumes that the particle size distribution remains constant with respect to time. At constant solids loading, a changing mean particle size results in a change in the BPI intensity, as the backscattered intensity is dependent on particle size. In a similar fashion, the BPI probe is well suited for determining changes in particle size as long as the solids loading remains constant.

CONCLUSIONS

A low-cost, in-situ fiber optic BPI probe has been developed which is well suited for measuring solids concentration and velocity in a gas-solid media. The probe design enables use in reactive-environments, thereby allowing *in-situ* analysis in many gas-solid unit operations. Although, for this study, the BPI was used to examined hydrodynamics near the wall, it can be used to examine radial variations in gas-solid flow by inserting the probe into the flow stream (providing intrusive effects are minimal).

The combination of a restricted field of view, using converging illuminating and collecting fibers, and the velocity modulus algorithm provides solids loading and velocity data in gas-solid streams with up to 50% solids loading. The range of sensitivity in the BPI technique (0 to >50% loading) makes it applicable to riser, standpipes, diplegs and pneumatic transport lines. Varying the sample frequency allows acquisition of slugging frequency and slugging size. In fluidized beds, the probe can provide data on solids loading, bubble frequency and bubble size.

A patent has been issued for this work [8] and the technology has been licensed to Perkin-Elmer Real-Time System Division.

REFERENCES

1. Halow, J.S., G.E. Fasching, P. Nicoletti and J.L. Spenik, Presented at AIChE Annual Meeting, Los Angeles, Nov 1991, Paper 105c.
2. Halow, J.S., *Process Tomography*, in press.
3. Halow, J.S. and P. Nicoletti, *Powder Technology*, **69**, 255 (1992).
4. Snell C.C., R.L. Dechene and R.E. Newton, *Instr. Chem. and Pet. Ind.*, **15**, 99 (1979).
5. Jotaki, T. and Y. Tomita, Pnuemotransport First Conference, Chirchill College, Cambrige (1971).
6. Maeda, M., K. Hishida and T. Furutani, Presented at the Symposium on Polyphase Flow and Transport Technology for ASME, San Fran., Aug 1980, p. 211.
7. Hollander, W., G. Pohlmann and G. Morawietz, *J. Aerosol Sci.*, **20**, 381 (1989).
8. R. Chrisman, R. Cocco and J. Cleveland, U.S. Patent #5365326, Nov. 15, 1994.
9. R. D. McLachlan and L. D. Rothman, U.S. Patent #4,707,134, Nov. 17, 1987,
10. J.W. Lyons, J.A. Roper III and P.A. Aldrich, U.S. Patent #4,978,863, Dec. 18,1990,
11. D. Geldart, "Gas Fluidization Technology," John Wiley & Sons, Chichester, (1989).
12. Louge, M., Presented at the 1994 AIChE Fall Meeting in San Francisco, CA, Nov, 1994, Paper 162d.
13. Haider, A. and O. Levenspiel, *Powder Technology*, **58**, 63 (1989).
14. Kerker, M., "The Scattering of Light and Other Electromagnetic Radiation," Academic Press, New York, (1969).

A Light/Charge Solids Flow Meter

James Rader and George E. Klinzing
Chemical/Petroleum Engineering Department, University of Pittsburgh, Pittsburgh, PA 15261

Anand Prakash
Viking Systems International, 2070 William Pitt Way, Pittsburgh, PA 15238

A new concept for the development of a novel solids flowmeter has been tested. This solids flowmeter is based on the principle of the interaction of the light field and flow of charged particles. The meter essentially utilizes a beam of polarized light that passes through an optical fiber wrapped around the outside of pipe. The electromagnetic field generated by charged particles flowing in the pipe induces a degree of rotation of polarized light. The particles were charged by bombardment with a nonconductive wall. Basic experiments conducted have proved the concept. The main components of the fiber-optic sensor assembled for the experiments were a laser source, single mode optical fiber, a fiber optic coupler, microscope objectives, beam splitter and detectors. The polarized laser beam was connected to the optical fiber with the help of a fiber-optic coupler. The output beam was detected for changes in solids flow rate. The signal from the sensor was very stable with standard deviation of les than 2%. The device was found to be sensitive to small changes in solids flow. This meter is expected to be reliable and durable with low maintenance cost since it is non-intrusive and has no moving part. This solids flowmeter will provide an accurate and instantaneous reading of the amount of solids flowing so that control can occur producing a more reliable operation.

INTRODUCTION

Pneumatic transport of solids is used in many commercial applications including coal utilization, food processing and mineral processing. The need for instrumentation in solids transport area is just as necessary as for those processes and operations that employ liquid and gases. However, the provision of a robust and reliable solids flowmeter is often difficult to achieve, whereas liquid or gas systems are more readily instrumented. The reliability of solids flow in comparison to that of gases of liquids has been estimated to be only 60% at best. This fact feeds directly to the overall system operation where solids are concerned. One large step in the right direction is to have a flowmeter that will provide an instantaneous reading of the amount of solids flowing so that better control can occur resulting in more reliable operation. Most industrial processes handling solids depend on the steady flow rate in order to have a reliable continuous operation. A flowmeter that functions on simple principles, and is non-intrusive into the flow would be a great asset to these solids processing operation.

Basically charged particles are made to flow through the test unit where the pipe is wrapped with an optical fiber. A polarized light source is inserted at one end of the fiber optic wire and the degree of rotation of the polarized light is measured in order to assess the amount of solids flowing throughout the pipeline. The flow of the charged particles induces a electromagnetic field on the region surrounding the flow and this induced field changes the degree of rotation of the polarized light. In doing this, one is able to obtain different readings of the degree of rotation depending on the concentration of the solids flowing inside the pipe.

When solids particles flow in pneumatic conveying they hit the wall and because of the differences in surface energy a charge is transferred often producing charged particles. If a section of the system is non-conductive then this charge that can be generated in pneumatic conveying is of the order of 0.1 coulombs/kg of material flowing. The charge of each particle usually varies from a large charge of 10^{-8} coulombs to 10^{-16} coulombs/particle for very small charges. Charging can vary depending on the quality of the solids being conveyed as well as the humidity of the carrier gas. These parameters can be controlled and thus reproducible charging can be achieved. It is the objective of every plant to produce solids that have tight specifications. One should also note that the electromagnetic field that is produced by the flowing charged particles varies with distance from the pipe wall. Changing pipe diameters and placement of the optical fiber in the field should be consistent since distance from the charged source will influence the strength of the field. With care this effect can be well managed.

OBJECTIVES

The two main objectives of this work were:

- To demonstrate the concept of interaction between the flow of charged- particles and polarized light using a fiber-optic sensor.

- To obtain preliminary data to show relationship between amount of solids flowing in a pipe and sensor response.

EXPERIMENTAL FACILITY

A common flow loop for pneumatic conveying was used to conduct the experiments. Figure 1 shows a schematic diagram of the gas-solid transport system used for this investigation. The major portion of the conveying loop consisted of 0.0509 m (2 in) internal diameter copper pipe. The total length was approximately 56 m connected with Morris couplings at some joints and soldered at others. Morris couplings were used for joining sections as these also maintain electrical conductivity for earthing. The pipe sections were electrically grounded at regular intervals. A 0.61 m (2 ft.) long 0.0508 m (2") I.D. transparent glass pipe provided a visual section for observation of the solids flow pattern and electrostatic charge developed in the horizontal plant. A 1.52 m (5 ft) long 0.0508 m (2") I.D. PVC pipe section was installed in the vertical section of the loop to measure electrostatic charge and electromagnetic field. Since solids flow is more homogeneous in the vertical section of the pipe, most of the measurements were made in this section of the loop.

A 20 HP Roots positive displacement blower provided the transport power. The blower is rated at 300 cfm at 10 psig. Suction and discharge silencers are provided for noise reduction. The blower was controlled by a local variable speed control panel. The rpm is changed by changing the frequency of the power supply. An orifice plate was used to measure the superficial gas velocity.

A 2 HP MSA screw feeder with variable pitch provided control of the solids flow from the storage bin to the transport line. The rpm of the motor could be varied by means of a variable frequency power supply which also compensated for loss of torque at low rpm.

Assembly and Testing of Fiber-Optic Device

The most challenging part of the project was putting together an optical device to measure the change in the state of polarization in response to solids flow. A photograph of the device assembled for the purpose is shown on Figure 2. The device consists of the following main components,

1. Laser
2. Polarizer
3. Fiber-optic coupler
4. Single Mode optical fiber
5. Microscope objectives
6. Quarter wave plate
7. Beam splitter
8. Photo detectors

Several problems had to be overcome before the device could be brought to working condition. One of the main problems was faced with connecting the laser into the single mode optical fiber which had a core diameter of about 4μm. Coupling laser light into the core of a single-mode fiber requires submicron resolution and stability. Moreover, the fiber end needs to be cut and polished to minimize scattering of laser beam. A high precision single-mode fiber coupler from Newport/Klinger was selected for this purpose. An input iris in the coupler assisted in centering the laser beam. The beam passed through a steering lens which is mounted in a precision X-Y translation assembly with 80 pitch adjustment screws. An objective lens focused the beam on the fiber core. A challenging problem was selecting a suitable optical-fiber for the device. The single-mode optical fiber selected for the device is of special design from 3M specialty fiber. The composition of the fiber made it sensitive to changes in the electromagnetic field strength. In addition, the fiber is twisted to minimize birefringence which may result from wrapping the fiber around the pipe. The fiber was wound 50 times around the pipe. Specifications of 3M fiber are given in Table 1. Ordinary optical fiber were not found suitable for the purpose. The output beam from optical fiber was collimated and passed through a quarter wave plate and a beam splitter to separate the two orthogonal linearly polarized modes. The two components-(I_1 & I_2) were detected with the help of photodetectors and the signals were sent to a computer for recording. The output response P is obtained by differencing and dividing the sum of the two components to stabilize against intensity fluctuation, i.e.

$$P = \frac{I_1 - I_2}{I_1 + I_2} \qquad (1)$$

This response is a function of change in the state of polarization of the light beam. Other configurations of the sensor system may also be possible to improve the response.

Table 1. Specifications of single mode optical fiber from 3M

Fiber type	FS-LB-3211
Fiber ID #	915922-8289
Attenuation	10.1 dB/km at 630 nm
ESI core radius	1.8 µm
ESI delta	0.32%
Num Aperture	0.12
Mode cutoff	570
Fiber diameter	80 ± 3 µm
Coating diam.	200 ± 15 µm

Principle of Operation of Fiber-Optic Sensor:
Light propagating through any medium possesses the property of polarization. The state of polarization (SOP) at any location refers to the behavior of the electric field vector E of the light wave as a function of time at that location (Azzam and Bashara, 1977). The most general SOP is elliptical polarization (ellipse becomes a circle) and linear polarization (ellipse collapses into a line). For any propagating wave, the SOP can be represented as the linear superposition of two waves having orthogonal SOPs, for example, two orthogonal linearly polarized waves or two circularly polarized waves having opposite senses of rotation (Udd, 1991). If the SOP changes in some way as light propagates through a material, the material is said to be birefringent. Birefringence can be classified according to the type of wave whose SOP is unaffected by the medium. Hence the SOP of circularly polarized light remains constant in a material having circular birefringence. The same material, however, has a strong effect on the SOP of linearly polarized light. In general, the total change in the SOP depends on the length of the light path in the birefringent material, the strength and type of birefringence, and the input SOP.

Faraday discovered that circular birefringence can be induced in a variety of materials by the application of a magnetic field. Therefore, it is possible to sense magnetic fields by using linearly polarized light to measure the amount of circular birefringence induced in certain materials by an external magnetic field. One can write an expression for the rotation of the plane of polarization of linearly polarized light as follows:

$$\theta = V \int B \cdot dl \qquad (2)$$

where the Verdet constant **V** is a measure of the field strength of the Faraday effect in the fiber and where the integral is performed over the length of the fiber exposed to the field magnetic field **B**. The Verdet constant depends on fiber composition and optical wavelength and is only weakly dependent on temperature in dielectric materials (Udd, 1991).

Birefringence in some materials change depending on the electric field that is applied. These electro-optic effects refer to the changes brought about in the light-refracting properties of a solid or liquid medium by the application of an electric field. The speed of components of light polarized parallel and perpendicular to the field is affected by the applied electric field. The Kerr effect, discovered in 1875, is the phenomenon where the change in refractive index is proportional to the square of the electric field. If the refractive index is linearly dependent on the field, the phenomenon is the Pockels effect. These effects can also be the basis for a solids flowmeter (see Appendix A for details).

Measurements with Fiber-Optic Sensor
The fiber-optic sensor for solids flow measurements consisted of three main points 1) Laser source 2) Optical fiber 3) Detection system. After initial testing and dry runs, measurements were made with solids flow.

Figure 3 shows a response curve for the flowmeter. The fiber-optic sensor responded well to increasing flow rate of solids. Following is a summary of observations made,

- The device responded well to changes in the flow rate of solids.
- There was also a good response to unsteady state conditions.
- The signal from the sensor was very stable with standard deviation of less than 2%.

Although the device responded well to changes in solids flow rate, the signal was susceptible to such disturbances as sudden electric discharge from test section. There is also the problem of residual charge after the flow was stopped. The test section on the conveying loop was not designed to eliminate these problems. The system, however, can be redesigned to minimize these problems.

Measurements of electrostatic potential around the test section indicated the presence of a strong electromagnetic field.

TECHNICAL AND ECONOMIC FEASIBILITY
Results from the experiments carried out for the project present us with an opportunity to develop a unique solids flowmeter. This flowmeter would have the following desirable characteristics,

- Durable
- Reliable
- No moving parts
- Low maintenance

Additional research is needed to make this device a commercial reality. Further work is being done to improve accuracy, sensitivity and drift of the instrument over a wide range of operating conditions. A dedicated microprocessor would allow signal processing and analyzing for the commercial instrument. The cost of development of this flowmeter is expected to be small fraction of the anticipated benefits from improved efficiency and reliability of solids flow achieved. The estimated cost of this solids flowmeter is about 20% lower than currently available meters for similar applications. Considering the current downward trend in the cost of the components of the flowmeter i.e. laser, optical fiber optics and electronics, the cost of commercial meter is expected to be even lower. Moreover operating cost of this meter is also expected to be very low due to low maintenance cost is absence of any moving parts.

LITERATURE CITED
Azzam, R.M.A. and Bashara, N.M. "Ellipsometry and Polarized Light", North-Holland, Amsterdam, p. 1-10 (1977).

Beach, R., Chem. Eng., Dec. 21, p. 73-78 (1964).

Cross, J.A., "Electrostatics - Principles, Problems, and Applications", Adam Hilger, Bristol, UK (1987).

Gibson, N., "Electrostatic Hazards," Institute of Physics (UK), Conf. Ser. No. 66, p. 1-11 (1983).

Longhurst, R.S., "Geometrical and Physical Optics", Longman (1973).

Maudlin, J.H. Light Lasers and Optics," TAB Books, Inc. Blue Ridge Summit, PA (1988).

Udd, E., Fiber Optic Sensors: An Introduction for Engineers and Scientists", John Wiley & Sons Inc. (1991).

Appendix - A

Brief Review of Scientific Principles

ELECTRIC AND MAGNETIC FIELDS

Considering the basic physics of light, one finds that light is made up of a combination of electromagnetic waves at right angles to one another. Therefore, like all electromagnetic radiation light is governed by the electromagnetic theory. Light, therefore can be influenced by the presence of a magnetic and/or electric field. Since the two fields propagate hand in hand; it is usually sufficient to consider either one and ignore the other. Utilizing polarized light, one finds that the degree of polarization can be affected by the strength of the electromagnetic field it is exposed to.

POLARIZED LIGHT

A wave may vibrate horizontally, vertically, or in any other direction; or it may vibrate in a complicated combination of horizontal and vertical oscillations. Such effects are called polarization effects. A wave that vibrates in a single plane (horizontal for example) is said to be plane-polarized. The polarization of ordinary light is considered random and has no preferred direction. If a piece of polarizing material is now inserted, the planes in which the electromagnetic waves travel can be controlled such that only a single plane exists. This can be clearly shown when two polarizers are used in conjunction with one another. If the transmitted axis (polarized plane) are parallel, the light will be transmitted. If they are perpendicular, no light will pass. By varying the angle between the transmission axes it is possible to control the intensity of the transmitted light.

MAGNET-OPTIC AND ELECTRO-OPTIC EFFECTS

The electro-optic and magneto-optic effects refer to the changes brought about in the light-refracting properties of a solid or liquid medium by the application of a magnetic and/or electric field across the medium. Utilizing polarized light, one finds that the degree of polarization can be affected by the strength of the electric or magnetic field applied. When charged solid particles flow through a pipe, the associated magnetic and electric field can be utilized to induce a degree of rotation of a polarized light that is confined to an optical fiber wrapped around the outside of the pipe. This concept is the basis for the solids flowmeter explored in this study.

FARADAY EFFECT

Faraday discovered in 1845 that when a block of glass was subjected to a magnetic field, it became optically active. When planes of polarized light are sent through glass in a direction parallel to the applied magnetic field, the plane of vibration is rotated. This phenomenon has been observed in gases, liquids and solids. One can write an expression for the rotation of the plane of polarization of linearly polarized light as follows:

$$\theta = V \int B \cdot dl \qquad (3)$$

where, θ - degree of rotation
              V - Verdet Constant in units of unit path/unit field strength
              B - field strength, teslas
              l - thickness of the path, m

Some Verdet Constants are given in Table 2.

TABLE 2 Verdet Constants for Several Mediums

MEDIUM	Temp, °C	V, arc min/tesla m
light		5893 micron
Water	20	1.31 E-4
Glass	18	1.61 E-4
Carbon disulfide	20	4.23 E-4
Ethanol	25	1.112 E-4
Quartz	20	1.66 E-4

It has been found that large effects can be obtained for light and electromagnetic interactions when employing films of ferromagnetic materials (Longhurst, 1973). For example, a value of Verdet constant of 0.0131 min of arc/oersted/cm has been measured for ferromagnetic films. The crystal YIG, yttrium-iron-granite is used in the infrared region where it is transparent and thus is useful in the application of the Faraday effect.

It should be noted that the refractive index of some materials change dependent on the electric field that is applied. The speed of components of light polarized parallel and perpendicular to the field is affected by the applied electric field. The Kerr effect, discovered in 1875, is the phenomenon where the change is refractive index is proportional to the square of the electric field while if the refractive index is linearly dependent on the field, the phenomenon is the Pockels effect. These effects could also be used as the basis for a solids flowmeter.

ELECTRO-OPTIC EFFECT

The electro-optic effect refers to the changes brought about in the light-refracting properties of a solid or liquid medium by the application of an electric field across the medium. In a demonstration of electro-optic effect, a beam of unpolarized light from a laser is first sent through a polarizer. The linearly polarized light that emerges has its electric field oscillating in the plane defined by the axis of polarization of the polarizer. This linearly polarized beam can be represented by its vertical and horizontal component which are in phase. As the two components pass through the electro-optic medium they are slowed down at different rates and hence gradually become out of phase. They emerge from the medium to form an elliptically polarized light beam. Another polarizer placed in the path of this beam, with its polarization axis oriented at 90 degrees with respect to that of the first polarizer, allows only the component that is parallel to its transmission axis to be transmitted. By varying the applied voltage the polarization of the beam can be made more or less elliptic, and as a result the amplitude of the output beam can be changed accordingly.

KERR EFFECT

When certain liquid and gases are placed in electric fields, their molecules tend to align themselves parallel to the direction of the electric field. The greater the field strength, the more complete the alignment of the molecules. Because the molecules are not symmetrical, the alignment causes the liquid to become anisotropic and birefringent. Such electric-field-induced birefringence in isotropic liquids is called the Kerr electro-optic effect or Kerr effect. The optic axis induced by the field is parallel to the direction of the field. For constant electric-field strength, the liquid behaves exactly as a birefringent crystal with indices n_o (the index of the material in the absence of the field) and n_e. The amount of field-induced birefringence ($n_o - n_e$) is proportional to the square of the electric-field strength and to the wavelength. Therefore, the index difference between the ordinary and extraordinary rays is,

$$\Delta n = K \lambda E^2 \qquad (4)$$

where K is the Kerr constant and E is the applied electric field. Nitrobenzene has an unusually large Kerr constant, $2.4 \times 10^{-23} cm/V^2$. Glasses have Kerr constants between about 3×10^{-14} and 2×10^{-23}. The Kerr constant of water is 4.4×10^{-12}.

POCKELS EFFECT

This is an electro-optic effect that is observed in certain crystals such as potassium dihydrogen phosphate (KDP). This effect differs from the Kerr effect in that the Pockels effect is linear in applied electric field, whereas the Kerr effect is quadratic in applied electric field. Moreover, the half-wave voltage of typical Pockels cells is at least an order of magnitude lower than that of the Kerr cell.

The Pockels effect causes a phase shift to occur in an optical wave propagating along the fiber in the presence of a dc field and is also a second-order nonlinear effect not normally found in fibers. The phase shift is linearly proportional to the applied electric field E_{dc} and may thus be used for linear electro-optic modulation.

FIBER OPTIC SENSORS

An optical fiber is basically a guidance system which is cylindrical in shape. If a beam of electromagnetic energy enters this system through one end-face of the cylinder, than a significant portion of this energy, usually light, will be trapped within the system and guided through it to emerge from the other end-face. The optical fiber has two concentric layers called the core and the cladding. The inner core is the light carrying part. The surrounding cladding provides the difference in refractive index that allows total internal reflection of light through the core. The index of the cladding is less than 1% lower than that of the core. Typical values, for example, are a core index of 1.47 and a cladding index of 1.46. Fiber manufacturers must carefully control this difference to obtain desired fiber characteristics.

Most fibers have an additional coating around the cladding. The coating, which is usually one or more layers of polymer, protects the core and cladding from shocks that might affect their optical or physical properties. The coating has not optical properties affecting the propagation of light within the fiber. This buffer coating, then, is a shock absorber.

The optical characteristics of fibers make it possible to develop sensors which can lead to technological improvements. Optical fibers present special characteristics of their own in addition to the qualities of optical sensors in general. The special advantages of optical fibers include absence of signal radiation outside the fiber, low weight, small dimensions, and great flexibility of the geometric configuration, ability to operate at high temperatures in corrosive of explosive environment, high sensitivity and chemical inertness.

Conventional optical fibers for the telecommunication purpose are designed and fabricated so as to be able to eliminate effects due to changes in the physical parameters such as the temperature, strain, pressure, rotation, electric field, magnetic field and moisture. Special fibers different from ordinary communication fibers are necessary for the high sensitivity measurement of physical parameters. Such special fibers can be obtained by performing structure modification and/or material system modification to the conventional fiber. The material system of the fiber can be modified at any part of the fiber structure such as the core, cladding, and coating with the special high sensitivity material which include glass, polymer, organic materials, metal and single crystal. In the structure change, fiber parameters can be changed using special structure in fiber cross section and in the fiber longitudinal direction. Higher sensitivity can be attained using special structures in the fiber cross section such as polarization maintaining structure, two mode structure, twin core structure, and evanescent waveguide structure. Higher sensitivity can be also attained with the special structure in the fiber longitudinal direction.

Physical parameters such as the temperature, strain, pressure, rotation, electric field, magnetic field and moisture are measured using the special fibers, by intensity modulation effects and phase modulation effects.

LASER LIGHT SOURCE

Light sources are today put into one of two categories, laser sources and thermal sources. Except with certain laser sources is said to be incoherent or partially coherent. Lasers emit light that is an electromagnetic wave with a well-defined frequency and phase. This coherence often results in a monochromatic, collimated output. Light with a well defined phase can be focused to a tiny spot to achieve very high intensities. Laser light source also allow measurement of phase of light in addition to intensity.

Although a laser is usually thought of as a coherent source, only a laser that oscillates in a single axial and transverse mode emits highly coherent light. A multimode laser, although much more intense, may be no more coherent (in space or time) than a suitably filtered thermal source.

PARTICLE CHARGING AND THE PHENOMENA

The generation of electrostatic charge in powder processing is essentially a separation of positive and negative charge. This charge separation is attributed to contact or frictional charging (known as triboelectrification), which occurs when materials of different electronic band structure (with differing electronic work functions) come into close mechanical contact. A listing of materials ordered according to their work functions is called a triboelectric series (Beach, 1964). At the boundary between two dissimilar materials, some of the electrons in the material with the lower work function are pulled across the interface and, when the material are separated, this charge imbalance results in a net positive charge for one body and a negative net charge for the other body. For example, a contact between clean metal and polymer surfaces usually accumulates positive charge in the metal and negative charge in the polymer. The physical separation of these bodies requires work, which becomes stored as electrostatic energy in the field.

A fundamental upper limit for particle charge can be derived based on the constraint that the electric field at the surface can not exceed the breakdown strength of air (i.e. E_b = 30 kV/cm). The upper limit upon charge per unit area σ_{max} (in Coulombs per square meter) is then

$$\sigma_{max} = \varepsilon_o E_b = 2.7 \times 10^{-5} \, C/m^2 \quad (5)$$

where $\varepsilon_o = 8.854 \cdot 10^{-12} F/m$ is the permitivity of free space. Based on this notion of a maximum surface charge, a limiting value for the specific charge $(q/m)_{max}$, also called the charge per unit mass, for spherical particle may be written,

$$(q/m)_{max} = 3\sigma_{max}/R\rho \quad (6)$$

where ρ is the mass density of the particle (in kg/m^3) and R is its radius (in meters). The limits predicted by Equation 6 are seldom achieved in practical situations. Typical charging levels sustained during pneumatic transport of powders are in the range 1 to $10^2 \mu C/kg$ (Gibson, 1983: Cross, 1987). The magnitude of the charging current depends on the following,

- Physical properties of the particles (shape, size, density)
- Conductivity, permitivity, humidity
- Pipe wall roughness, Pipe diameter and length traversed by particles
- Solids velocity and solids concentration

The net charge acquired and maintained during gas-solid flow in an isolated section of the transport system causes the formation of an electric field in the isolated section. The electric field can be written as a function of the charge generation and transfer in a general form according Gauss's Law.

$$\varepsilon_o \int (E \cdot dA) = q \quad (7)$$

For a constant electric field parallel tot he Gaussian surface containing a net positive charge, the above equation can be written as:

$$E = q/(\varepsilon_o A) \quad (8)$$

The electric field generated in the transport system can easily by measured as a function of the potential difference across the pipe wall using an electrostatic voltmeter. The following equation shows the relationship.

$$V_{diff} = V_f - V_i = -\int (E \cdot dS) \quad (9)$$

For a constant electric field parallel and in the same direction as the Gaussian path, the above equation can be written as:

$$E = V_{diff} / s \quad (10)$$

By knowing the electric field for a given flow rate it may be possible to develop an optical fiber device using the Kerr effect.

Conductive Induction

The term "conductive induction" describes the process by which an initially uncharged particle that comes into contact with a charged surface assumes the polarity and eventually, the potential of the surface. A particle of a good electrical conductor will assume the polarity and potential of the charged surface very rapidly. However, a dielectric particle will become polarized so that the side of the particle away from the charged surface develops the same polarity as the surface. Particles of intermediate conductivity may be initially polarized but approach the potential of the charged surface at a rate depending on their conductivity.

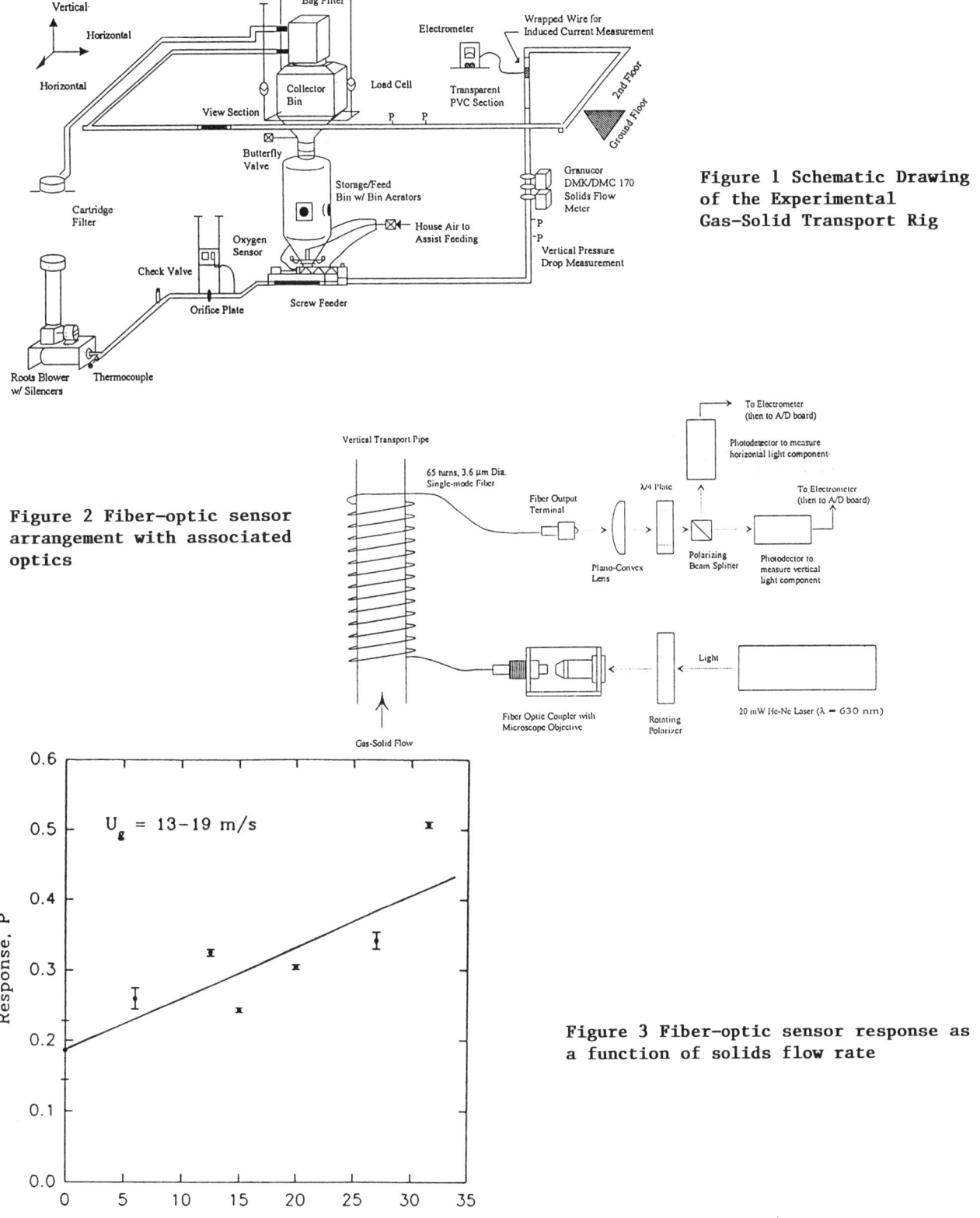

Figure 1 Schematic Drawing of the Experimental Gas-Solid Transport Rig

Figure 2 Fiber-optic sensor arrangement with associated optics

Figure 3 Fiber-optic sensor response as a function of solids flow rate

Non-Intrusive Measurement of Solids/Liquid Concentration and Velocity: Use of MRI Visualization

J.X. Bouillard
Argonne National Laboratory, 9700 South Cass Avenue, Argonne, IL 60439

and

S. Altobelli
Lovelace Medical Foundation, 2425 Ridgecrest Drive, Albuquerque, NM 87108

Cross-sectional two-phase liquid/solids slurry flow patterns in a horizontal rotary mixer are measured using Magnetic Resonance Imaging (MRI) flow visualization techniques. The liquid/solids mixture consists of 3300 microndiameter glass beads dispersed in an aqueous solution. The mixer is a horizontal cylinder which can rotate at various angular speeds. The use of a non-intrusive velocity MRI phase encoding technique revealed complex three-dimensional flow patterns in both the dense slurry phase and the surnatant pure liquid phase. This technique shows great potential in capturing slurry mixing transport mechanisms in rotary mixer in a non-intrusive manner.

Introduction

The flow of solids-loaded suspension has been the subject of intense experimental and theoretical investigations. These types of flows are of great interest to chemical engineers because of their importance in many industrial manufacturing processes. Such flows are encountered, for example, in the manufacture of solid rocket propellants, advanced ceramics and reinforced polymer composites; They are also encountered in heterogeneous catalytic reactors and in the pipeline transport of liquid-solids suspensions. Slurry processes similar to those using tomato juices and paper pulp are lso frequently encountered in the food and paper industries.

In most cases, the suspension microstructure and the degree of solids dispersion greatly affect the final performance of the manufactured product. For example, solid propellant pellets need to be extremely well distributed in polymer matrices so that they can be used as solid fuels for rocket engine. The homogeneity of pellet dispersion is critical for the uniformity of the burn rate, which, in turn, affects the final mechanical performance of the engine. Commonly used noninvasive flow measurements are usually based on light (laser doppler anenometry), sound waves, X/gamma rays, and particle beam or electrical impedance techniques, which all suffer from attenuation, opacity, and scattering across immiscible-phase interfaces, especially at high solids loadings.

In contrast, Nuclear Magnetic Resonance (NMR) techniques usually do not experience transmission effects in nonconducting media and are truly noninvasive. NMR imaging presents the potential to noninvasively determine hydrodynamic parameters of complex flow processes, such as solids/liquid volume fractions and multiple-phase-flow velocities [1, 2, 3, 4, 5, 6, 7].

Because of the recent developments in pulsed Fourier Transform NMR imaging, NMR imaging is now becoming a powerful technique for the nonintrusive investigation of multi phase flows. Kose et al. [8] originally introduced two-dimensional NMR flow-imaging techniques which result in both velocity profiles and flow-compensated liquid-volume fraction distributions.

Rotating cylinders are used in chemical and metallurgical industries for processes as comminuting, mixing, ball milling and drying. Most of the granular or slurry flow studies in rotating cylinders have been performed by direct visual observation techniques. In this paper, we present a new non-intrusive measurement technique that can apply to complex multiphase flow suspensions, such as those encountered in slurry mixers.

This paper reviews the experimental set-up used in MRI visualization, namely the magnet, the RF and gradient coils and the Fourier reconstruction techniques. These techniques are then applied to investigate flow patterns in liquid/solids rotary mixers.

Experimental Methodology

The rotary mixer is made of an acrylic cylinder of 10 cm long by 6 cm diameter. The cylinder is filled by a mixture of water and glass beads at 35 % solids volumetric fraction. The glass beads settle down in the lower part of the cylinder at a high solids concentration (i.e 65 %) while the upper part of the cylinder is filled with water only. The glass beads are 3300 micron in diameter and have a specific solids density of 2.64 g/cm^3. The cylinder is allowed to rotate azimuthally at various speeds, thereby mixing the solids with the liquid phase. The mixer is set in motion by a 3 m long non-magnetic shaft mounted on a speed-controlled electrical motor. The motor is a dc servo motor (12 FG) which is controlled by a controller VXA-48-8-8, made by PML.

Solids concentration and velocity measurements were performed in the horizontal bore (31 cm diameter) of a superconducting helium cooled magnet (1.98 T from Oxford Instruments). A "bird-cage" radio frequency (rf) probe, tuned to 80.3 MHz [2], and actively shielded gradient coils (Magnex) were controlled by a versatile VAX-based (Digital Equipment Corp.) imager/spectrometer (Quest 4400 from Nalorac Cryogenics Corp.). Data collected by the NMR imager was transferred via ethernet to a workstation (Sun Microsystems) for analysis using customized data-reduction software. A schematic representation of the MRI facility is shown in Figure 1. NMR imaging experiments make use of the proportionality between NMR frequency and magnetic field expressed in the Larmor equation $\frac{\omega}{2\pi} = \gamma B$ where γ has the value 42.5 MHz/T for protons. Gradient coils were used to generate spatially linear variation of B of the form $B_o + G_x x + G_y y + G_z z$, which translates into a frequency mapping of the three-dimensional space. The gradients were small compared to the static field: the peak values for the gradients used in these experiments were 1-4 mT/m. NMR signals were induced by the small, transient, magnetization resulting from the precession of a large number of nuclear spins. This magnetization was initially excited by an rf pulse, modified by gradient pulses, and then observed in the presence of a "readout" gradient. The process encoded one spatial coordinate in the frequency of the signal and additional spatial (and velocity) information in the phase of the signal. After observation, the magnetization was allowed time to reequilibrate.

The use of NMR imaging for fluid velocity measurements is relatively new but it is gaining acceptance as the method is improved. Most previous measurements have been taken for simple flow geometries, i.e., unidirectional flows, where application of the NMR technique is more straightforward [2,7]. The difficulty in doing NMR of complex flows is the irreversible dephasing which occurs during the time of measurement; this is caused by incoherent motions within an image element (voxel). We have succeeded in making velocity measurements in the liquid/solid rotary mixer for fairly slow flows by using short time intervals between pulses. In this work, we used a standard spin-warp rf pulsing sequence combined with the phase method for velocity determination [2]. Spin echo experiments were performed to reduce the sensitivity of the NMR sequence to overall magnetic field inhomogeneities. Standard spin echo proton concentration measurements with velocity compensation were use to measure the three-dimensional distributions of solids concentrations.

The timing sequences of the rf pulse, the magnetic field gradients, and the resulting NMR signal are shown in Figure 2. The amplitude-modulated rf pulse marked $\pi/2$ and the first lobe of G_s defined a 10 mm thick slice perpendicular to the axis of the mixer. Thus, each NMR image was of a cylindrical volume 6 cm in diameter and 1.0 cm in axial length. The π pulse produced a spin-echo 12 ms after the center of the $\pi/2$ pulse. $G_{r(x\ or\ y)}$ encoded one transverse coordinate in the frequency of the signal and $G_{p(x\ or\ y)}$ imparted a phase shift depending upon the other transverse displacement. The readout gradient G_r provided velocity sensitivity when the dotted waveform was followed. For each NMR image produced with this method, N_2 or (128) repetitions of the basic sequence, differing in the value of G_p, were collected. For each of these repetitions, which were separated by 300 ms, N_1 or (128) points were digitized during the period indicated by the horizontal arrow. The N_1 by N_2 complex array thus obtained was made into an image by a two-dimensional discrete Fourier transform. The voltage profile of the read-out gradient G_r controlled the sensitivity of the NMR signal to the r-component (x or y) of velocity. Two profiles are shown in the figure: the solid line shows a balanced, or velocity-insensitive waveform and the dotted line shows a velocity-sensitive profile. The final lobe of G_r was present in both waveforms. Images are made with both profiles and v_r is calculated from the phase difference of the two images. In these experiments this procedure was performed with both horizontal and vertical readout gradients. Logically, the stepped gradient G_p can occur at any time in the NMR sequence. For velocity measurements, however, it was important to place the phase-encoding pulses near the time of echo formation to reduce registration errors which occurred because the read-out and phase-encode coordinates were encoded at slightly different times. Ideally, the G profiles have velocity sensitivity in proportion to the first moment $\int tGdt$. Because of experimental imperfections, mostly eddy currents induced in the metallic cryostat by the rapid changing magnetic field gradients, there were small spurious contributions to the image phase which we canceled with "extra" reference images. Mathematical division of the complex data was used rather than subtraction of phases because it reduced the number of phase wraps in the velocity image. In these experiments, the flow velocity and sensitivity were large enough to cause phase wrapping (velocity induced phase shifts greater than π) in regions of relatively high velocity (as near the cylinder wall). Spatial regions where this occurred were bounded by large discontinuities because the calculated phase jumped between branches of the Arg function. The Arg function is the function used to compute the arctangent of the quotient of the imaginary and real images. Points near the tube wall were adjusted by adding multiples of 2π to the phase in the wrapped regions.

Results and Discussion

Based on the spin-echo technique discussed above, and using the velocity-insensitive pulsing sequence described above, an image of the solids concentration can be visualized in the two-dimensional slice as displayed in Figure 3. A dynamic angle of repose of about 20 degrees can be measured from this image. As can be seen, this technique is powerful to display non-intrusively the two-dimensional distribution of solid particle volume fractions.

In the lower part of the mixer, the solids volume fraction nears 65 %, and an area of low mixing can be identified. In this area, the solids tumble down and do no mix with the upper liquid/solids layers. In the supernatant phase, the liquid is mainly entrained by the motion of the rigid walls of the cylinder. The cylinder motion induces a liquid velocity distribution at the cylinder wall which propagates inward toward the core of the mixer by viscous effects. Cross sectional velocity components along horizontal and vertical median cutting lines of the device are shown in Figures 4 and 5. As can be seen in Figures 4 and 5, the velocity boundary layers are significantly altered by the presence of particles. The rotating cylinder tangential velocity was set to 9 cm/s. As illustrated in Figures 4 and 5, the maximum velocity measured near the cylinder wall approaches 6 cm/s, which is indicative of the difficulty in accurately resolving the velocity profile at the cylinder wall. This difficulty to measure velocity profiles in wall regions with NMR methods stem from a) magnetic field inhomogeneities that are present at these interfaces and b) the phase wrapping phenomenon discussed above due to existence of high velocities in this particular region.

As the liquid phase is entrained and accelerated from points A to B, along the peripheral arc of the mixer (See Figure 3), the liquid phase comes into contact with the dense slurry phase at point B. Because of blockage effects due to the presence of solids particles which locally increase viscous drag forces upon the liquid phase, the liquid does not significantly penetrate the dense slurry but follows a path of least resistance by moving upward along the line BC. At the same time, a downward liquid flux moves in the direction of line AC, resulting from particles tumbling down and dragging the liquid along. At point C (in Figure 3), these two liquid fluxes converge and are forced upward toward the top of the mixer (i.e following the path of least resistance). At this low rotating velocity, the surnatant liquid phase exhibits two large scale vortices as indicated in Figure 3. As the angular rotation velocity of the mixer nearly doubles, these two vortices merge to form a single large vortex as indicated in Figure 6. A mathematical model for the two-phase flow of this mixer is presently being developed using the liquid-slurry flow software developed by FLUCOMP Inc.

Conclusions

Three-dimensional measurements of solids concentrations and liquid velocities were performed in a rotary mixer using non-intrusive MRI visualization techniques. It is clear that such technique can provide information important to the design of larger and more efficient mixers. As reported in this paper, only the protons of the aqueous phase were imaged. This technique could be easily extended for two nuclei, as for example the proton of water and fluorine (^{19}F) that could be inserted into the beads. By using a two-nuclei MRI technique, imaging of volume fractions and velocity fields for both liquid and solids phases could be performed simultaneously.

Acknowledgments

This work was supported by the U.S. Department of Energy under the contract W-31-109-Eng-38.

References

[1] P. D. Majors, R. C. Gilver, and E. Fukushima. Velocity and concentration measurements in multiphase flows by NMR. *J. Magn. Reson.*, 855:235–242, (1989).

[2] A. Caprihan and E. Fukushima. Flow measurement by NMR. *Phys. Rep.*, 198:195–235, (1990).

[3] S. A. Altobelli, A. Caprihan, and E. Fukushima. *NMR Flow Studies by Phase Methods*. San Francisco Press, San Francisco, 1990.

[4] S.A. Altobelli, R. C. Gilver, and E. Fukushima. Velocity and concentration measurements of suspensions by NMR Imaging. *J. Rheology*, 35(5):721–734, (1991).

[5] S. W. Sinton, J. H. Iwamiya, and A. W. Chow. NMR imaging of industrial flow processes. *Mat. Res. Soc. Symp. Proc.*, 217:73–78, (1991).

[6] S. W. Sinton and A. W. Chow. NMR flow imaging of fluids and solid suspensions in Poiseuille flows. *J. Rheology*, 35(5):735–772, (1991).

[7] J. X. Bouillard. Nmr imaging and hydrodynamic analysis of neutrally buoyant non-newtonian slurry flows. *Powder Technology*, 78:99–103, (1994).

[8] Kose K., K. Satoh, T. Inouye, and H. Yasuoka. NMR flow imaging. *J. Phys. Soc. Jpn*, 54:81–92, (1985).

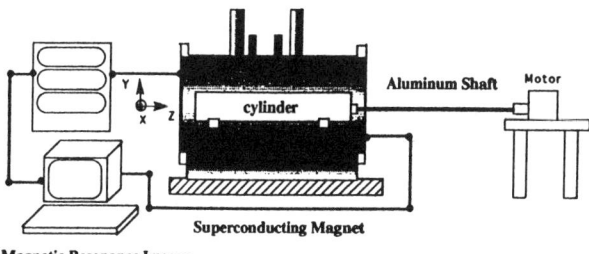

Figure 1: Schematic View of the Experimental Apparatus with the Magnetic Resonance Imager, a 1.9 T superconducting magnet and a motor located about 3 m away from the magnet to rotate the mixer

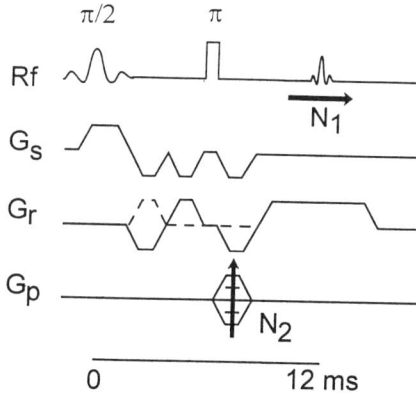

Figure 2: Spin Warp and Phase-Encoded Velocity NMR Pulse sequence

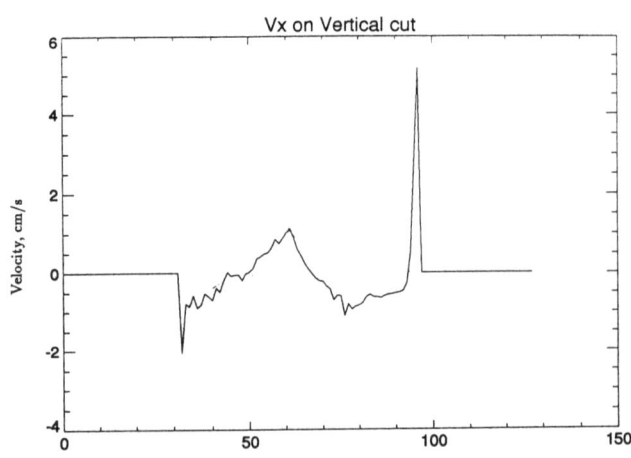

Figure 4: Horizontal Velocity Profile Along the Median Vertical Plane

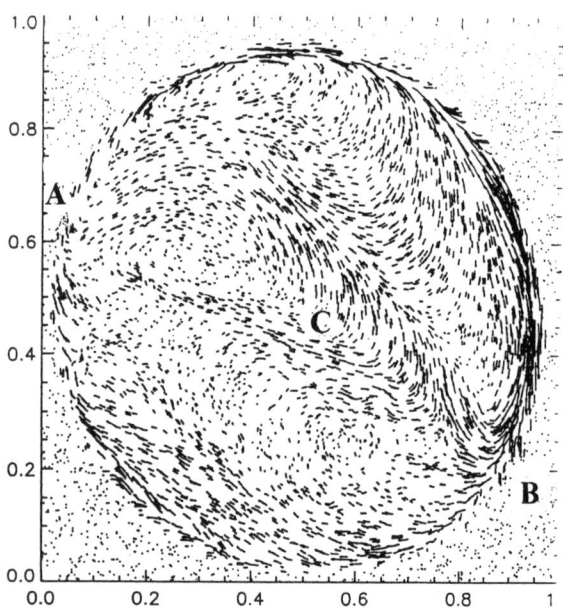

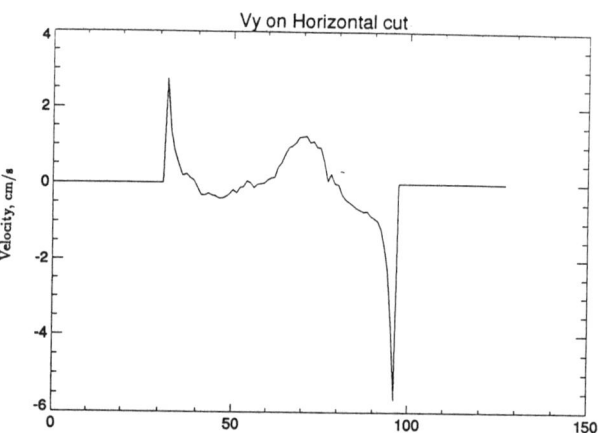

Figure 5: Vertical Velocity Profile Along the Median Horizontal Plane

Figure 3: Liquid/Solids Mixing in a Rotary Mixer: Top; Non-Intrusive MRI Two-Dimensional Measurement of Solids Concentration, Bottom: Two Dimensional Liquid Velocity Map (Rotation Velocity 9 cm/s)

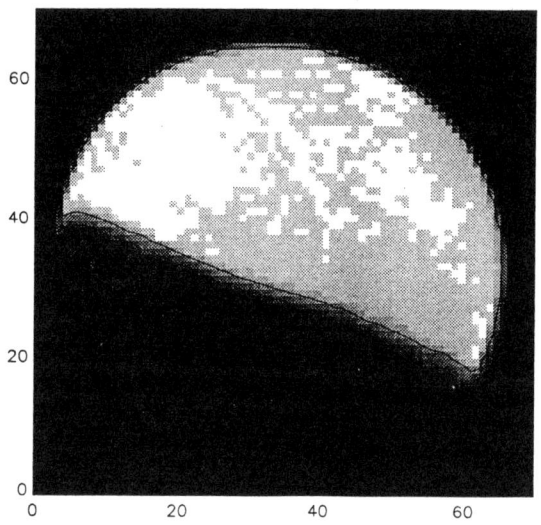

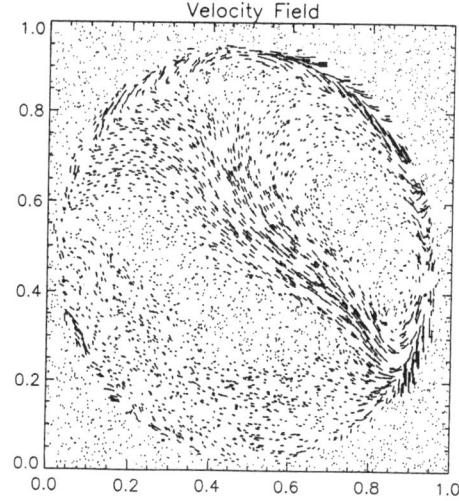

Figure 6: Liquid/Solids Mixing in a Rotary Mixer: Top; Non-Intrusive MRI Two-Dimensional Measurement of Solids Concentration, Bottom: Two Dimensional Liquid Velocity Map (Rotation Velocity 18 cm/s)

Characterization of Fluidization Properties of Fine Powder FCC Catalyst at Elevated Temperatures
- Prediction of Rheological Parameters -

H.O. Kono, Y. Itani, E. Aksoy, E. Koresawa, and J.J. Su
Department of Chemical Engineering, West Virginia University,
Morgantown, WV 26506

To characterize quantitatively fine powder fluidization at elevated temperatures (25-600 °C), the values of rheological parameters were found to be much more useful than using classical particle data of particle size, its density, and shape factor. Two rheological parameters, namely the plastic deformation coefficient (Y) and fracture strength (σ_f) of the powder structure were previously defined and experimentally measured at ambient temperaure under the aerated conditions without any bubbles, i.e., at gas velocities between the minimum fluidization and the minimum bubbling points (Kono, et. al., 1994). In this study, the measurement of the powder rheological parameters was extended to elevated temperatures (25-600 °C). The activation energy of plastic deformation coefficient of FCC catalyst was found to be approximately 8.6 Kcal/Kmol, so that rheological parameter Y can be predicted at elevated temperature from that at ambient temperature.

In the past literature, the effect of temperature on the fluidization characteristics has been mostly interpreted in terms of the change of the fluidizing gas properties by temperature [1,2,3]. There has also been reported discrepancies about the evaluation of minimum fluidization velocity, showing that U_{mf} of relatively fine powder behaves distinctively different from that of coarse particles. It should be due to the voidage change of the emulsion phase for coarse particle ($d_p > 100\mu$) at high temperature as reported by Rowe [4]. However, the temperature effect on the surface property change of fine fluidized particles ($d_p < 70\mu$, e.g., FCC catalyst, acrylonitrile catalyst, etc.) has not really received due attention in spite of its industrial significance. This seems to be simply because there has been no reliable experimental measurement method available to characterize rheological parameters of the emulsion phase at elevated temperatures under the uniformly aerated condition.

Although the approach was completely different from this paper, Jimbo et al. [5] proposed a pioneering empirical approach to predict the minimum fluidization velocity at high temperatures, taking the particle interaction force under the packed bed condition into consideration. His proposed equation claimed to include the change of the adhesion force with respect to temperature. Unfortunately his application of the split cell tester, however, introduced some disadvantages such as: (1) that technique could not be utilized directly to characterize the aerated powder properties. But it could only be applicable to the relatively dense packed powders, and (2) the accuracy of the split cell method is not satisfactory to exactly measure the low tensile strength of powders such as FCC, or other fluidizable catalysts.

The characterization of the flow property of fine powders has been frequently requested to understand the various powder handling processes such as transportation, mixing, flow conditioning, fluidization, etc. Rietema [6] first indicated the importance of the emulsion phase quality in fluidized beds, taking the effect of interparticle forces on fluidization into consideration. Kono et al. [7] and Khoe et al. [8] measured the tensile strength of dry fine powders used for fluidization under a certain packed bed condition and utilized it to correlate with the properties of the emulsion phase of the fluidized beds. The powder packing structures in the processes of mixing, transportation, fluidization, etc., are pretty close to the aerated powder packing condition, which is significantly different from the densely packed bed condition. Most of the industrial fine powders with the size of 30 to 70 μm, e.g., FCC catalysts, have very low tensile strengths of the powder structure under the operating conditions. Based upon the above considerations, a measurement method of rheological parameters for the aerated fine powders was developed by combining the principles of fine powder aeration and powder rheology.

A novel measurement method was developed in this work to define the rheological properties of the uniformly aerated fine powders at elevated temperatures. Two important rheological parameters, the plastic deformation coefficient (Y) and fracture strength (σ_f) of the packing structure of fine powders under the aerated condition were theoretically defined and experimentally measured, using spent FCC at elevated temperatures (25 to 600 °C).

Based upon this study, we found several important facts:

Firstly, this method was found to be very useful to quantitatively measure these rheological parameters of fluidizable fine powders.

Secondly, the parameter σ_f could be used as an index of bulk powder flowability.

Thirdly, the parameter Y could be also used as a volume expansion-index of the aerated powders. The activation energy of Y was found as 8.6 Kcal/Kmol by Arrhenius plotting.

Fourthly, there is an interesting relationship between these two rheological parameters: i.e., Y and σ_f could be expressed approximately as $\sigma_f = 0.11\ Y^{0.70}$, which holds true for all the experimental data of sample fine powders regardless of their kinds and properties. It seems that this Y-σ_f relation is a necessary condition for fine powders to be fluidized, although further study should follow.

Fifthly, when the spent FCC powder was used at elevated temperature, Y and σ_f values became both smaller. The set of Y and σ_f values at elevated temperature were still on the same characteristic line of σ_f-Y described above. To attain better fluidization, the rheological parameters should be as small as possible and at the same time the sets of Y and σ_f should be on the characteristic line of Y-σ_f. The general tendency is that the higher the temperature, the better the quality of fluidization.

Sixthly, the ratio of σ_f/Y was found to be crucial to maintain a satisfactory fluidization quality on the Y-σ_f characteristic line. Generally speaking, when the experimental results are derailed from that characteristic line, agglomeration or defluidization will happen.

Finally, another useful utilization of these rheological parameters is that there is a clear qualitative relationship between rheological parameters and visual bubble images obtained in two dimensional fluidized beds.

THEORETICAL CONSIDERATION

These key rheological parameters of fine powders determining fluidization characteristics were defined in the gas velocity range of minimum fluidization, U_{mf}, and minimum bubbling points, U_{mb}. The derivation of the rheological parameters was described briefly in the attached Appendix. The aerated powder bed is assumed to consist of N uniform powder layers as shown in Figure 1a, and N is defined as:

$$N = \frac{H_{mf}}{d_p} \qquad (1)$$

Within the range of the gas velocity of $u_{mf} < u < u_{mb}$, when the gas velocity increases, the powder layer expands uniformly and homogeneously. At U_{mb} point, a fracture occurs, ending the homogeneous powder structure and causing the first bubble appearance. This is our interpretation that the first formation of the bubbles should simultaneously cause the first fracture of the homogeneous aerated powder emulsion. This interpretation was justified by the observation of bubble formation together with the bed expansion in two and three dimensional fluidized bed. It is experimentally verified that the first bubble occurring at u_{mb} is not always located near the gas distributor but also at the upper portion of fluidized beds.

Between U_{mf} and U_{mb}, the voidage of the powder bed changes from ϵ_{mf} (the minimum fluidization voidage) to ϵ_{mb} (the minimum bubbling voidage) as shown in Figure 1c, while the bed pressure drop, ΔP_D, stays constant. However, if the voidage change would not be allowed between U_{mf} and U_{mb}, the pressure drop for the whole bed would reach a value of ΔP_H, where the difference ($\Delta P_H - \Delta P_D$) was called as the excess pressure drop, $\Delta P^*(U_{mb})$, as shown in Figure 1b. Considering that the homogeneously aerated bed structure consists of N uniform layers, $\Delta P^*(U_{mb})$ per layer, i.e., $\Delta P^{**}(U_{mb})$, was derived by utilizing the viscous term of Ergun equation as:

$$\Delta P^{**}(U_{mb}) = \frac{\Delta P^*(U_{mb})}{N} = 150 d_p \frac{(1-\epsilon_{mf})^2}{\epsilon_{mf}^3} \frac{\mu}{(\phi_s d_p)^2} (U_{mb} - U_{mf}) \qquad (2)$$

The effect of inertia term of Ergun's equation was found to be negligible small within the range of this study. From the equality of the energy necessary to expand the bed supplied by the gas flow per unit time and the energy consumed by the deformation of powder structure for this expansion per unit time, the fracture strength at the minimum bubbling point $\sigma_{f,mb}$ was derived, using the "principle of virtual work" (i.e., $E_D = E_\sigma$) as shown in Figure 2b and 2c. The physical meaning of E_σ is that E_σ is the amount of energy per unit time to deform the homogeneous emulsion. E_σ is being supplied by E_D. The fracture strength at minimum bubbling point $\sigma_{f,mb}$ can be derived as Equation 3:

$$\sigma_{f,mb} = \frac{\Delta P^{**}}{\left[1+\left(\frac{\varepsilon_{mb}}{\varepsilon_{mf}}\right)\left(\frac{1-\varepsilon_{mf}}{1-\varepsilon_{mb}}\right)\right]} \quad (3)$$

The strain of the deformation (S) can be written as Equation (4):

$$S = \frac{(h_{mb}-h_{mf})}{h_{mf}} = \frac{N(h_{mb}-h_{mf})}{N\, h_{mf}} = \frac{H_{mb}-H_{mf}}{H_{mf}} \quad (4)$$

where h_{mf} and h_{mb} show the height of the unit layer at minimum fluidization and minimum bubbling point respectively, as shown in Figure 3. The plastic deformation coefficient Y can be written as Equation (5):

$$Y = \frac{\frac{\Delta P^*}{N}}{\left(\frac{h_{mb}-h_{mf}}{h_{mf}}\right)} = \frac{\Delta P^{**}}{\left(\frac{H_{mb}-H_{mf}}{H_{mf}}\right)} \quad (5)$$

The details of equations of 3, 4, and 5 are to be referred to Appendix. Although we used various equations to show the physical meaning, it is very important to note at this point that both of the rheological parameters, the fracture strength (σ_f) and the plastic deformation coefficient (Y), were obtained simply by using experimental data. The advantage of using these theological parameters is not only to describe the aeration characteristics quantitatively but also to understand the fluidization flow properties in terms of its physical meaning.

The traditional way of powder characterization by size, shape and density did not provide any insight to the emulsion property in terms of physical meaning. Contrary to that, the rheological parameters can represent the aerated powder's structure and the deformation properties.

Therefore, the effect of temperature on the rheological parameter, e.g., Y can be expressed by using the concept of activation energy, as will be discussed more in the following section of the experimental results and discussion. However, it is to be noted that this fact indicates that the aerated powder structure did change with the temperature. This type of understanding could be reached only through this proposed approach.

EXPERIMENTAL

The experimental parameters, i.e., ΔP^{**}, U_{mb}, U_{mf}, H_{mb} and H_{mf}, that are necessary to obtain the fracture strength at the minimum bubbling point ($\sigma_{f,mb}$) and plastic deformation coefficient (Y) were obtained through the well-known collapse test as described by Rietema [6], Abrahamsen et al. [9] and Kono et al [7].

The experimental apparatus is schematically shown in Figure 4. The cylindrical fluidized bed was 10.2 cm in diameter, which is appropriate to avoid the wall effect on collapse tests for the sample powders. The gas distributor was a sintered porous plate with an average opening size of 20 μm. The fluidizing gas was supplied from a dry high pressure air line via flowmeter. A video recorder was used to record collapse experiments. The minimum bubbling points were measured in the same way as reported. The experimental apparatus was also equipped with electrical heaters. As a result, all the experimental parameters, i.e., ΔP^{**}, U_{mb}, U_{mf}, H_{mb} and H_{mf}, were obtained from the experimental data. These data were utilized to calculate the rheological system parameters by using the equations given in the previous section. As sample powders, spent FCC, glass beads, special starch-A and B, and carbon black were used for the tests at ambient temperature and only spent FCC was used for the tests at elevated temperature. The relevant physical properties of these powders are given in Table 1.

The overall and local bubble images of freely bubbling fluidized beds were also recorded by using a video recorder (image time intervals of 1/30 s) to visualize the bubbling characteristics of the emulsion phase. As an experimental equipment, a two-dimensional fluidized bed (25 cm x 150 cm x 1 cm) was employed and the powders were fluidized at the gas velocity at 9 cm/sec, using five different kinds of fine powders.

This measurement method should be applied to fine powders, where u_{mb} is significantly larger than u_{mf}.

RESULTS AND DISCUSSIONS

The experimental results of rheological

parameters, Y and $\sigma_{f,mb}$ at ambient temperature are shown in Figure 5a. The fracture strength $\sigma_{f,mb}$ of the aerated powders represents approximately the fracture strength of the emulsion phase in fluidized beds. All the sample powders are selected to be fluidizable. At the same time, all the sample powder selected does not form any agglomeration, although some particles are very fine. In other words, this method is most effective to characterize quantitatively the fluidizable powders with the relatively low interparticle forces, e.g., various industrial catalyst powder, starch, toner powders for printing, etc. Within the range of this study, the range of $\sigma_{f,mb}$ of the powders at ambient temperature under the aerated condition was from 0.005 to 0.3 Pa, which is generally too small to be measured accurately by the conventional split-cell method. The value of σ_f of FCC is in the range of 0.07-0.3 Pa under the temperature range of 25-600 °C.

Certainly, it is interesting to find that there is a clear quantitative correlation between these two rheological parameters of σ_f and Y for all of our five fluidizable sample fine powders. Regardless of their kinds and properties of these sample fine powders, the Y and $\sigma_{f,mb}$ values were approximately on identical one line regardless of the ranges of particle size, density and surface characteristics within the acceptable range of experimental scattering.

The results provided an empirical equation as equation (6).

$$\sigma_{f,mb} = 0.10 \, Y^{0.70} \qquad (6)$$

Historically, the question of how the fluidization quality can be expressed was the matter of keen discussion since the 1960's. According to Geldart's classification diagram (1973), the starch-A powder ($d_p=15$ μm, $\rho_p=1550$ kg/m³) should belong to Type C powders and should not be fluidized. However the experiment results showed that this starch-A could beautifully be fluidized. The experimental evidence indicated that the particle size and density are not enough to predict the quality of fluidization. The experimental set of Y and σ_f of starch-A is located on the same characteristic line expressed in Figure 5a. In Figure 5b, the σ_f-Y experimental data at elevated temperature were shown, changing the temperature in the range of 25 to 600 °C. It is interesting that the σ_f-Y characteristic curves stay still on the same identical one characteristic line. In the case at elevated temperature, the higher the temperature, the smaller the values of σ_f and Y, and the better the quality of fluidization for FCC catalyst.

For all the experimental results shown in Figures 5a and 5b of σ_f and Y for various fluidizable powders with different properties and size (shown in Table 1) at ambient and elevated temperatures, the very characteristic line of Y and σ_f remains identical. Only the common property of all the sample powder at various temperatures is that they are all fluidizable and do not form agglomeration nor segregation. Although further justification should follow, the characteristic function of σ_f and Y ($\sigma_{f,mb}=0.1Y^{0.7}$) may represent the necessary condition for fluidization.

The bed expansion ratio of H_{mb}/H_{mf} can be correlated to Y as shown in Figure 6. The more the bed expands, the smaller the value of Y, where the values of Y and σ_f satisfy Equation. (6). By using Figure 6, the prediction of the aerated powder expansion behavior with a specific powder property can to some extent quantitatively be accomplished, showing an example of usefulness of rheological parameters.

From the temperature dependency of Y, the activation energy of the plastic deformation coefficient of aerated FCC powder was obtained by using Arrhenius plotting as shown in Figure 7. The activation energy seems to be approximately 8.6 Kcal/Kmol, assuming the molecular weight of FCC catalyst as 223. The physical meaning of this result indicates that the aerated powder structures get softer at elevated temperature.

In order to visually show the effect of these rheological parameters on fluidization, the bubble image recording in an appropriate two-dimensional fluidized bed seems to be the simplest and probably the most effective method, although the experimental limitation caused by the wall effect of two-dimensional bed on the bubbling phenomena should also be remembered.

In order to assess the effect of the change of rheological parameters on the bubble images in two-dimensional fluidized beds, Photo-picture 1 is shown. Even though at this point it is only qualitatively in terms of bubble images, it indicated that the change of rheological parameters provided the significant effect on bubble images in two-dimensional fluidized beds.

CONCLUSIONS

1. A novel measurement method was developed

to define the rheological properties of the aerated fine powders, which is useful to characterize the flow property of fluidizable powders quantitatively. These two rheological parameters; the plastic deformation coefficient (Y) and fracture strength (σ_f) of the aerated powders were theoretically defined and experimentally measured by using spent FCC, glass beads, carbon black and special starch-A and B at ambient temperature and using only spent FCC at elevated temperature.

2. Both of the rheological parameters; Y and $\sigma_{f,mb}$ at ambient and elevated temperature are found to be useful to predict the flow behavior of fine powders such as Y for the prediction of the aerated powder expendability and σ_f for characterizing the flow index.

3. There is a clear experimental relationship between rheological parameters of Y and σ_f for all the fluidizable sample fine powders regardless the powder properties and size at ambient and elevated temperatures, which can be expressed as $\sigma_f = 0.11\ Y^{0.70}$. The ratio of σ_f/Y was found to be crucially important to attain a good fluidization quality.

4. The activation energy of the plastic deformation coefficient (Y) for FCC powder was found to be approximately 8.6 Kcal/Kmol. Therefore, Y at arbitrary temperature can be predicted.

5. Even though it is still qualitative, it seems that there is certainly an interesting correspondence between these rheological parameters and freely bubbling characteristics in two-dimensional fluidized beds. The higher the temperature, the smaller the bubble size.

ACKNOWLEDGEMENT

This research was partially supported by Asahi Chemical Industry Co., Ltd., and with respect to the bubble image analysis, it was partially supported by U.S. Department of Energy-Morgantown Energy Technology Center, DE-AC21-92MC29222.

NOTATION

d_p	Average particle diameter, (μ)
E_D	Expansion energy defined by Equation (7A), (Pa.m/s)
E_σ	Energy necessary for expansion defined by Equation (6A), (Pa.m/s)
F	Interparticle force at one single contact point, (N)
H	Bed height, (m)
h	Height of the unit layer of powder bed, (-)
k	Coordination number of particle packing, (-)
N	Number of the unit layer of powder bed, (-)
ΔP	Pressure drop of the bed, (Pa)
ΔP^*	Excess pressure drop defined by Equation (3A), (Pa)
ΔP^{**}	Excess pressure drop per unit layer defined by Equation (5A), (Pa)
U	Superficial gas velocity, (m/s)
Y	Plastic deformation coefficient of the powder structure, (Pa)

Greek letters

ϵ	Voidage of the bed, (-)
μ	Viscosity of the gas, (Pa.s)
ρ	Density, (kg/m^3)
ϕ	Carman's shape factor of the particle, (-)
σ_f	Fracture strength of the powder structure, (Pa)
σ_t	Tensile strength of the powder structure, (Pa)

Subscripts

m	Bed height including bubbles
mf	At minimum fluidization
mb	At minimum bubbling

LITERATURE CITED

1. MATHUR, A., S.C. SAXENA and Z. F. ZHANG, *Powder technol.*, **47**, 247 (1986)

2. WU, S.Y. and J. BAEYENS, *Powder Tech.*, **67**, (1991)

3. RASO, G., M. D'AMORE, B. FORMISANI and P.G. LINGNOLI, *Powder Tech.*, **72**, 71 (1992)

4. ROWE, P.N., *Chem. Eng. Sci.*, **42**, 387 (1987)

5. JIMBO, G., R. YAMAZAKI, J. TSUBAKI and H. KAMIYA, Memoirs of the Faculty of Engineering, Nagoya Univ., Japan, (1988).

6. RIETEMA, K., in A. Drinkenburg (ed.), *Proc. Int. Symp. on Fluidization*, The Netherlands University Press, Amsterdam, 154 (1967).

7. KONO, H.O., in Transport Phenomena in Fluidized Particle Systems, ed. L.K. Doraiswamy and A.S. Mujumdar, Elsevier Science Publishers, B.V. Amsterdam (1989).

8. KHOE, G.K., T.L. IP and J.R. GRACE, Powder Technology., **66**, 127 (1991).

9. ABRAHAMSEN, A.R. and D. GELDART, *Powder*

Technol., **26**, 47 (1980)

10. Kono, H. O., Ediz Aksoy and Yoshihito Itani, *Powder Technol.* **81**, 177 (1994)

APPENDIX: DERIVATIONS OF RHEOLOGICAL PARAMETERS

Fine powders can be fluidized without forming any bubbles in the gas velocity range of U_{mf} and U_{mb} (fine powder under this condition is called hereafter as "the aerated fine powders"). In this range, the emulsion phase can expand uniformly without the influence of gravitational force. The reasonable assumptions in the following discussions are that the aerated powders should form a homogeneous, uniform structure consisting of many layers of fine powders as schematically shown in Figure 1a, in the gas velocity range of $U_{mf} \leq U \leq U_{mb}$. This model prevails under the defined aerated condition while the powder packing structure expands homogeneously. Finally, at the velocity of U_{mb}, the first fracture occurs, ending the homogeneity of the powder packing structure, and leading to the appearance of bubbles. This is a very important definition of the initiation of the bubble formation correlating itself to the first initiation of the fracture of the aerated powder's homogeneous structure. It is to be noted that the bubble characteristics, e.g., bubble size, shape, etc., which are the outcomes of the emulsion fracture, can be determined by the method described here.

The following physical changes occur in a bed of fine powder in the range of the defined aerated condition, i.e., $U_{mf} \leq U \leq U_{mb}$, (Figures 1b and 1c). Note that, the pressure drop across the bed, i.e., ΔP_D, actually remains constant in this range due to the increase of the bed voidage (Figure 1b). By Ergun's equation's first term, ΔP_D can be expressed as:

$$\Delta P_D = 150 H_{mf} \frac{(1-\varepsilon_{mf})^2}{\varepsilon_{mf}^3} \frac{\mu U_{mf}}{(\phi_s d_p)^2} = 150 H_{mb} \frac{(1-\varepsilon_{mb})^2}{\varepsilon_{mb}^3} \frac{\mu U_{mb}}{(\phi_s d_p)^2} \quad (1A)$$

On the contrary, if the bed height would remain at the same as H_{mf}, i.e., ε_{mf} would be constant, during the increase of the gas velocity from U_{mf} to U_{mb}, then the pressure drop across the bed, ΔP_H, at the minimum bubbling point would be (Figure 2b):

$$\Delta P_H = 150 H_{mf} \frac{(1-\varepsilon_{mf})^2}{\varepsilon_{mf}^3} \frac{\mu U_{mb}}{(\phi_s d_p)^2} \quad (2A)$$

For this hypothetical case, the difference of ($\Delta P_H - \Delta P_D$) was defined here as the excess pressure drop ΔP^*, as shown in Figure 2b. Therefore, at U_{mb}:

$$\Delta P^*(U_{mb}) = \Delta P_H(U_{mb}) - \Delta P_D = 150 H_{mf} \frac{(1-\varepsilon_{mf})^2}{\varepsilon_{mf}^3} \frac{\mu}{(\phi_s d_p)^2}(U_{mb}-U_{mf}) \quad (3A)$$

Although ΔP^* can be expressed as Equation (3A), still this value can directly be measured by experiments.

The total number of unit layers, N, can be defined in terms of the particle mean diameter (d_p) as:

$$N = \frac{H_{mf}}{d_p} \quad (4A)$$

Thus, from Equations (3A) and (4A), the excess pressure drop per unit layer at U_{mb} can be derived as:

$$\Delta P^{**}(U_{mb}) = \frac{\Delta P^*(U_{mb})}{N} = 150 d_p \frac{(1-\varepsilon_{mf})^2}{\varepsilon_{mf}^3} \frac{\mu}{(\phi_s d_p)^2}(U_{mb}-U_{mf}) \quad (5A)$$

In Figures 2b and 2c, the excessive pressure drop per unit powder layer and the fracture strength were plotted as a function of the gas velocity. The fracture strength of aerated powders (σ_f) can be expressed by Rumpf's equation:

$$\sigma_f = \frac{1-\varepsilon}{\pi} k \frac{F}{d_p^2} \approx \frac{1-\varepsilon}{\varepsilon} \frac{F}{d_p^2}$$

In Figure 3, on the other hand, the expansion process is schematically shown at U_{mf} and U_{mb} conditions. The energy necessary to expand the unit layer from h_{mf} to h_{mb} per unit time is:

$$E_a = \int_{u_{mf}}^{u_{mb}} \sigma_f(U) \, dU \quad (6A)$$

Similarly, the source of this expansion energy per unit time is:

$$E_D = \int_{u_{mf}}^{u_{mb}} \Delta P^{**}(U) \, dU \quad (7A)$$

Based upon the energy balance and applying the principle of virtual work to Equations (6A) and (7A), Equation (8A) can be derived:

$$E_D = E_\sigma \qquad (8A)$$

In Figure 2c, if one assumes that $\sigma_f(U)$ is linear between U_{mf} and U_{mb}, then:

$$E_\sigma = \frac{1}{2}(\sigma_{f,mb} + \sigma_{f,mf})(U_{mb} - U_{mf}) \qquad (9A)$$

On the other hand, from the definition of ΔP^{**}, i.e., Equation (5A) at any U, E_D can be integrated to give:

$$E_D = \frac{1}{2}\Delta P^{**}(U_{mb})(U_{mb} - U_{mf}) \qquad (10A)$$

Using Rumpf's simplified equation, the ratio of $\sigma_{f,mb}$ and $\sigma_{f,mf}$ can be expressed as a function of ϵ_{mb} and ϵ_{mf}. Accordingly,

$$\sigma_{f,mb} = \frac{\Delta P^{**}}{\left[1 + \left(\frac{\epsilon_{mb}}{\epsilon_{mf}}\right)\left(\frac{1-\epsilon_{mf}}{1-\epsilon_{mb}}\right)\right]} \qquad (11A)$$

$$Y = \frac{\frac{\Delta P^*}{N}}{\left(\frac{h_{mb}-h_{mf}}{h_{mf}}\right)} = \frac{\Delta P^{**}}{\left(\frac{H_{mb}-H_{mf}}{H_{mf}}\right)} \qquad (12A)$$

The physical meaning of $\sigma_{f,mb}$ expressed by Equation (11A) is the fracture strength of the aerated powder at u_{mb}, where the first fracture initiates within the uniform emulsion phase, generating the bubble initiation. The physical meaning of Y is clearly shown in Equation (12A). Y is the ratio of stress and strain, indicating the plastic deformation coefficient of the emulsion phase in the range of u_{mf} and u_{mb}.

It is very important to note at this point that, in this study, both the fracture strength and the plastic deformation coefficient results were obtained only by using experimental data. The detail of this appendix is to be referred to Kono et al. [10].

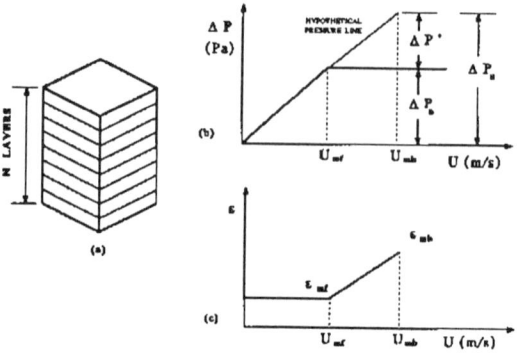

Figure 1. The physical changes occurring between U_{mf} and U_{mb} in a fine powder bed. (a) N layer representation of the powder structure, (b) The pressure drop vs. gas velocity, (c) The bed voidage vs. gas velocity.

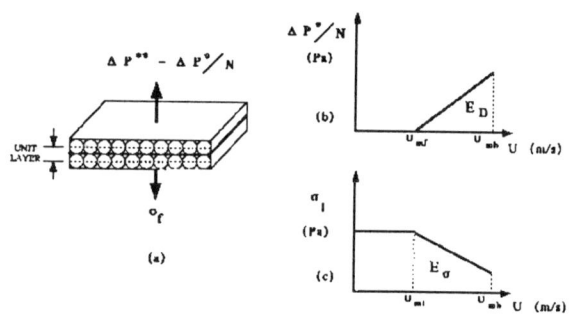

Figure 2. (a) The unit layer representation of the powder structure, (b) The excessive pressure drop per layer vs. the gas velocity, (c) The fracture strength vs. gas velocity.

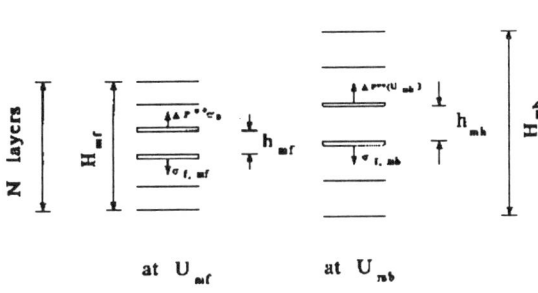

Figure 3. Schematic representation of the powder layer's expansion at U_{mf} and U_{mb} conditions.

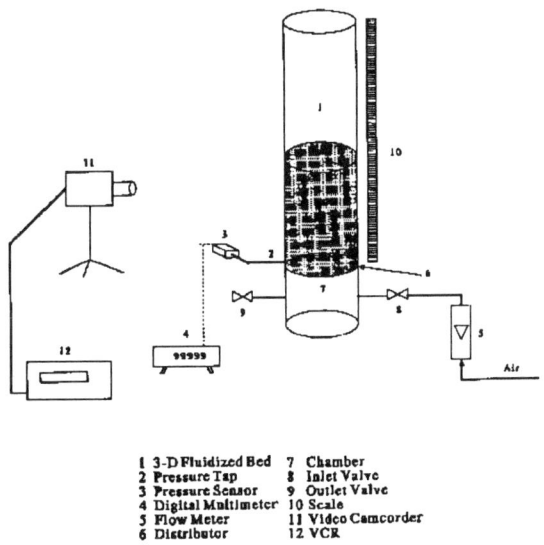

1 3-D Fluidized Bed 7 Chamber
2 Pressure Tap 8 Inlet Valve
3 Pressure Sensor 9 Outlet Valve
4 Digital Multimeter 10 Scale
5 Flow Meter 11 Video Camcorder
6 Distributor 12 VCR

Figure 4. Schematic diagram of experimental apparatus for the collapse test.

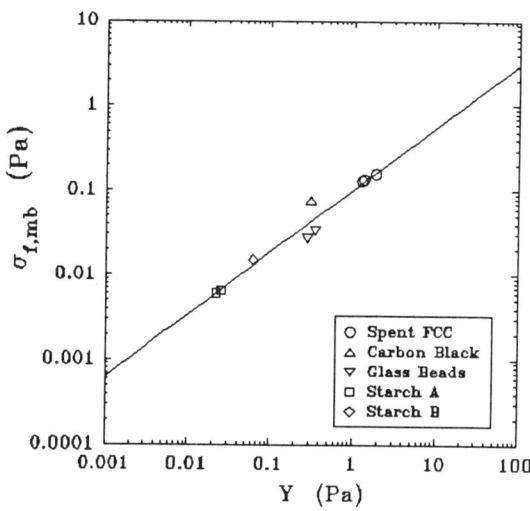

Figure 5a. Fracture strength $\sigma_{f,mb}$ vs. plastic deformation coefficient Y for fine sample powders at ambient temperature ($\sigma_{f,mb}=0.10Y^{0.7}$).

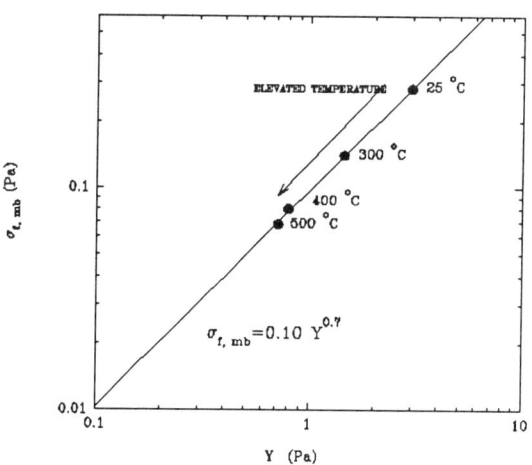

Figure 5b. The effect of temperature on the rheological properties (Y and $\sigma_{f,mb}$ of spent FCC)

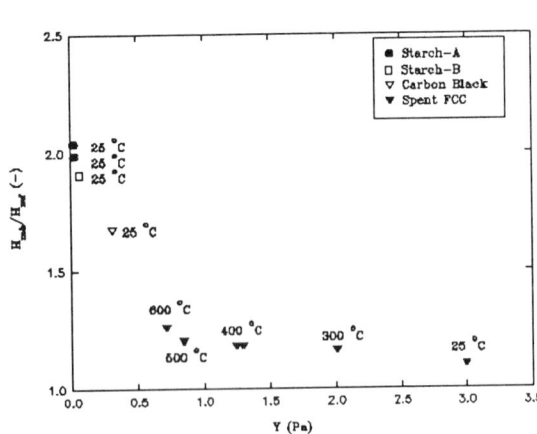

Table 1. The physical properties of sample powders.

	Powder	d_p μm	ρ_p kg/m^3
1	Spent FCC	55, 60, 65 (20 < d_p < 125)	1850
2	Carbon Black	30 (20 < d_p < 50)	1800
3	Glass Beads	30 (10 < d_p < 60)	2500
4	Starch-B	40 (30 < d_p < 50)	1391
5	Starch-A	15 (10 < d_p < 20)	1550

Figure 6. The relation between the bed expansion ratio and the plastic deformation coefficient for the sample powders at ambient and elevated temperatures.

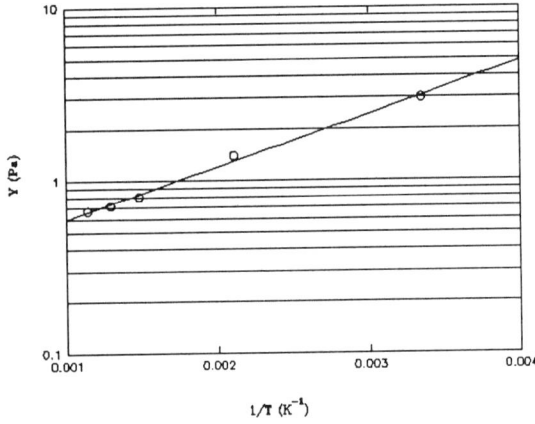

Figure 7. Arrhenius plotting of Y and T of FCC powder

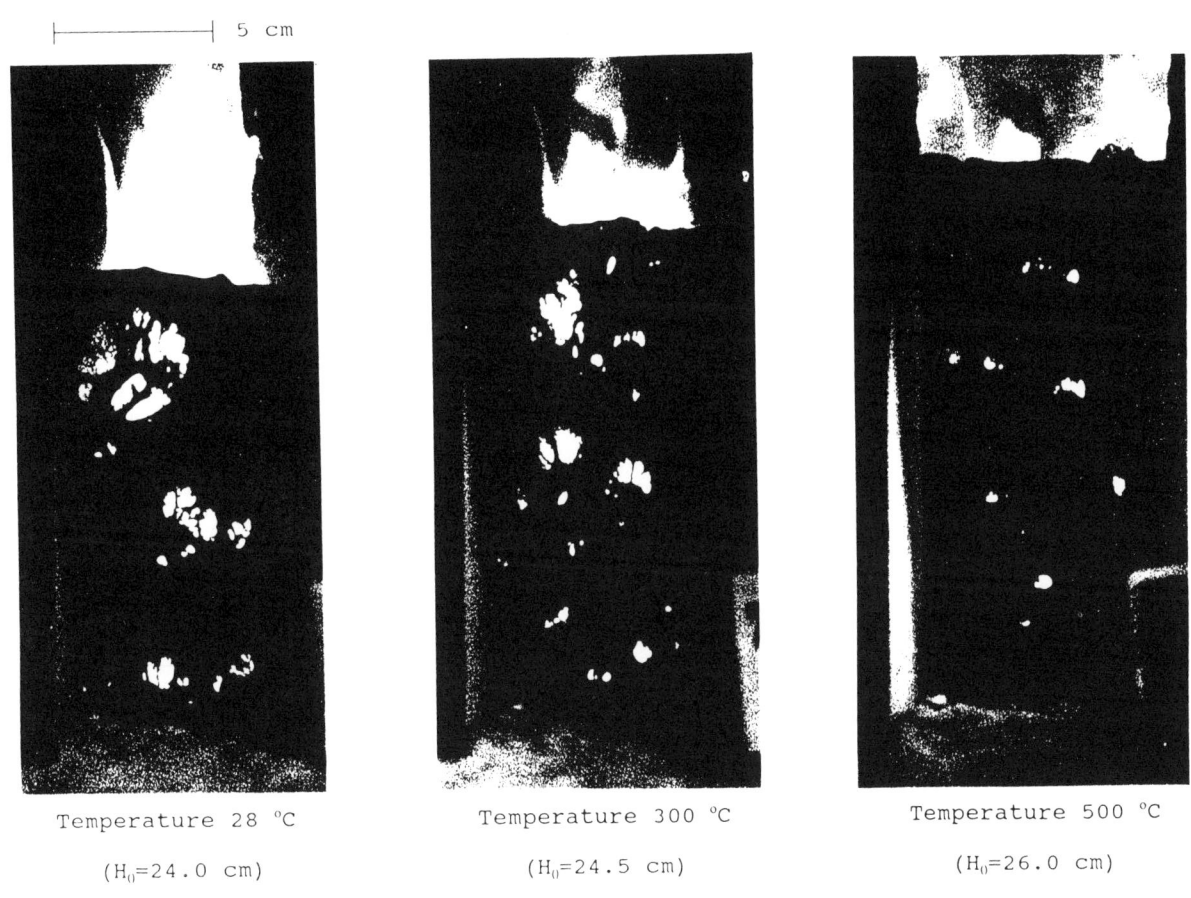

Picture 1 Bubble Images in Fluidized Beds at Various Temperatures

Spent FCC was used as fluidized particles. The gas velocity at each specific temperature was maintained constant at 9.0 cm/s. The bed height before fluidization was 21 cm.

Index

A

Axial gas dispersion 118,127

B

Bottom bed and transport disengagement heights .. 92
Bubble dynamics in liquid-solids suspensions 1

C

Catalyst, FCC 92
Channel, fluid-particle mixture in a 111
Chloride, hydrogen 70
Chlorine, converting hydrogen chloride to 70
Circulating fluidized beds 102
Coal conversion processes 81
Conveying, dilute phase pneumatic 136

D

Desulfurization, regenerative 81
Dilute phase pneumatic conveying 136

F

FCC Catalyst
    beds of fresh 92
    fine powder 169
Flow meter, solids 153
Flow rate monitoring and measurement 136
Fluid-bed scale-up 118,127
Fluid-particle mixture, gravity flow of 111
Fluidization properties, characterization of 169
Fluidized Bed(s)
    circulating 102
    description of 44
    granulators 60
    process 70
    system 81
    reactor 51

G

Gaseous medium, velocities in a 146
Gas residence time distribution 118
Granulation processes 60
Granulators, fluidized bed 60
Gravity flow of fluid-particle mixture 111

H

Hydrodynamic attractors 102

Hydrodynamics, influence of 81
Hydrogen chloride 70

I

Interconnected fluidized bed system 81
In-situ determination of particle loadings 146

L

Light/charge solids flow meter 153
Liquid-solids suspensions 1

M

Microreactor 118,127
MRI visualization, use of 163

N

Non-steady-state reaction kinetics 127

P

Particle loadings, *in situ* determination of 146
Pneumatic conveying, dilute phase 136
Pressure fluctuations, using 136
Process, fluidized bed 70

R

Reactor, fluidized bed 51
Regenerative desulfurization 81
Rheological parameters, prediction of 169

S

Solid mixing 127
Solids flow meter 153
Solids/liquid concentration 163

T

Transport disengagement heights, bottom bed and . 92
Turbulent fluid bed 118,127

V

Velocities in a gaseous medium 146
Velocity, solids/liquid concentration and 163
Vibrated bed ("vibrofluidization") 118,127
Voidage perturbations, infinitesimal and finite 44